AF873680

Using Network and Mobile Technology to Bridge Formal and Informal Learning

Chandos
Learning and Teaching Series

Series Editors: Professor Chenicheri Sid Nair and Dr Patricie Mertova
(emails: sid.nair@uwa.edu.au and patricie.mertova@education.ox.ac.uk)

This series of books is aimed at practitioners in the higher education quality arena. This includes academics, managers and leaders involved in higher education quality, as well as those involved in the design and administration of questionnaires, surveys and courses. Designed as a resource to complement the understanding of issues relating to student feedback, books in this series will respond to these issues with practical applications. If you would like a full listing of current and forthcoming titles, please visit our website, www.chandospublishing.com, email wp@woodheadpublishing.com or telephone +44 (0) 1223 399140.

New authors: we are always pleased to receive ideas for new titles; if you would like to write a book for Chandos, please contact Dr Glyn Jones on gjones@chandospublishing.com or telephone +44 (0) 1993 848726.

Bulk orders: some organisations buy a number of copies of our books. If you are interested in doing this, we would be pleased to discuss a discount. Please email wp@woodheadpublishing.com or telephone +44 (0) 1223 499140.

Using Network and Mobile Technology to Bridge Formal and Informal Learning

Edited by
Guglielmo Trentin
and
Manuela Repetto

CP
CHANDOS
PUBLISHING

Oxford Cambridge New Delhi

Chandos Publishing
Hexagon House
Avenue 4
Station Lane
Witney
Oxford OX28 4BN
UK
Tel: +44 (0) 1993 848726
E-mail: info@chandospublishing.com
www.chandospublishing.com
www.chandospublishingonline.com

Chandos Publishing is an imprint of Woodhead Publishing Limited

Woodhead Publishing Limited
80 High Street
Sawston, Cambridge CB22 3HJ
UK
Tel: +44 (0) 1223 499140
Fax: +44 (0) 1223 832819
www.woodheadpublishing.com

First published in 2013

ISBN: 978-1-84334-699-9 (print)
ISBN: 978-1-78063-362-6 (online)

© The editors and contributors, 2013

British Library Cataloguing-in-Publication Data.
A catalogue record for this book is available from the British Library.

All rights reserved. No part of this publication may be reproduced, stored in or introduced into a retrieval system, or transmitted, in any form, or by any means (electronic, mechanical, photocopying, recording or otherwise) without the prior written permission of the Publishers. This publication may not be lent, resold, hired out or otherwise disposed of by way of trade in any form of binding or cover other than that in which it is published without the prior consent of the Publishers. Any person who does any unauthorised act in relation to this publication may be liable to criminal prosecution and civil claims for damages.

The Publishers make no representation, express or implied, with regard to the accuracy of the information contained in this publication and cannot accept any legal responsibility or liability for any errors or omissions.

The material contained in this publication constitutes general guidelines only and does not represent to be advice on any particular matter. No reader or purchaser should act on the basis of material contained in this publication without first taking professional advice appropriate to their particular circumstances. All screenshots in this publication are the copyright of the website owner(s), unless indicated otherwise.

Typeset by Domex e-Data Pvt. Ltd, India
Printed in the UK and USA.

Contents

List of figures and tables

Figures

Tables

List of abbreviations

BYOD:	bring your own device
COP:	community of practice
CTE:	continuing teacher education
HE:	higher education
ICT:	information and communication technology
M-learning:	mobile learning
NEET:	not in education, employment or training
NMT:	network and mobile technology
PAH:	pedagogy-andragogy-heutagogy
PLE:	personal learning environment
SDT:	self-determination theory
SNS:	social network site
VLE:	virtual learning environment
WMD:	wireless mobile device

Preface

> In this Age of Mobilism, fueled by the technology, learning and schooling will undergo the most dramatic and disruptive change in the past 150 years – since the invention of the one-room schoolhouse, in fact. (Norris and Soloway, 2011)

No discussion about the integration of communication technology into teaching and learning processes can ignore the wide and indeed ever-widening gap between the students' personal/daily use of Web 2.0 and mobile technology and the manner in which schools propose using them for educational activities: on the one hand, we have the newest generation of students (sometimes called 'digital natives') with their innate capacity for mastering technologies, using them for both social and learning purposes, albeit with little or no method; on the other hand, there are the teachers ('digital migrants') who, while making admirable attempts to innovate their teaching, tend to propose an educational use of 2.0 resources based on conventional methods and practices of study, anchored to old teaching schemes. The ideal, however, would be to conceive and introduce new methodologies inspired by the so-called *e-pedagogy*, which are able to exploit fully the potential of network technologies both for social interaction and for accessing information.

E-pedagogy presents an innovative approach to teaching/learning processes, characterised by informal processes (among both students and teachers) aimed to answer specific, localised knowledge needs, in which serendipity is often synonymous with incidental learning.

But as this type of learning is not planned by either student or teacher, the problem of its connection to formal learning paths arises. In this sense we need to understand how much of this 'activism', which is frequently immediate and instinctive and expresses itself through videos on YouTube and posts and links on social networks, might be transferred to formal learning contexts and adapted to formal curricula.

In other words, it is worth considering how the uses of network and mobile technology (NMT) which have been autonomously acquired by students and teachers outside the school context could be channelled towards new educational paradigms which combine formal and informal learning processes centred on accessing information and social interaction.

With this in mind, the aim of this book is to analyse and discuss: (a) the conditions that should ensure real sustainability in the integration of informal and formal/institutional learning dimensions; and (b) the potentialities that NMT offers to support such integration.

The book is divided into eight chapters, each of which offers a complementary point of view on how NMTs can help bridge the gap between formal and informal learning.

In Chapter 1, Cathy Tran, Mark Warschauer and AnneMarie Conley demonstrate how educators are looking for ways to tap the educational potential of mobile technology as it becomes increasingly prevalent. The chapter argues that one of the biggest strengths of handheld mobile devices is their potential for increasing learning motivation. Elements such as their small size, geolocation awareness, instant information access and tactile features all contribute to a portable and personal experience that can motivate learners.

The theme of motivation is also addressed by John Cook, Norbert Pachler and Ben Bachmair, who in Chapter 2 highlight the potential of social media and personally-owned mobile devices for bridging the gap between media use in everyday life and the requirements of school and higher education. The chapter offers a wide perspective over access to 'cultural resources' facilitated by digital media (e.g. learning resources, health information, cultural events, employment opportunities, etc.). The purpose of the chapter is to reclaim the notion of 'cultural resources' for the educational environment as it seems relevant to discussions about social mobility and fairness of access.

In Chapter 3, Thomas Cochrane and Roger Bateman present a mobile Web 2.0 framework for pedagogical change, based upon the implementation of over 30 mobile learning projects between 2006 and 2011. The projects utilised mobile devices for student-generated content and for enabling student-generated learning contexts. The framework is based on contemporary social learning theory and illustrates the potential of mobile Web 2.0 tools to bridge the gap between formal and informal pedagogically-designed learning contexts and to cross international boundaries.

In Chapter 4, Julie Willems and Debra Bateman discuss the use of Facebook in higher education. Using a knowledge-sharing framework created by the intersection of two continua (formal and informal learning opportunities, and student-directed and teacher-centred learning), and an exemplary Facebook case study for each of the four quadrants of the framework, the chapter explores institutional (or teacher)-driven informal teaching and learning; student (or member)-driven informal teaching and learning; student (or member)-driven formal learning; and institutional (or teacher)-driven formal learning.

Also in the school context there is a strong need to understand how digital media can best be exploited to innovate teaching, making the media correspond more closely to students' communicative habits. However, many studies demonstrate how current school organisation still constitutes a barrier for any development related to the integrated use of digital media. The interconnections and networking between these two life-worlds experienced by young people present a challenge to researchers. To investigate such networking, Solveig Roth and Ola Erstad have used a learning ecology methodology to identify those explicit physical or virtual contexts that can provide opportunities for learning in everyday learning life. In Chapter 5, the two authors describe their methodology, pointing out how interest-driven learning, based on out-of-school digital practices, enables students to use research-based methods for methodological reasoning and structured problem-solving and to draw on interests and expertise from different contexts.

Two conditions are required to achieve these aims: (a) an organisational development of the school institution to foster didactic innovation in which NMT use is the normal condition, just as use of the traditional blackboard, paper and pen were for a long time the normal condition; and (b) a process for the continual professional development of teachers which is no longer (or not only) based on formal training so much as on informal learning processes specifically centred on NMT use.

In Chapter 6, Guglielmo Trentin discusses the complementary nature of these two conditions. Organisational development aimed at pedagogical innovation cannot work without the corresponding professional development of teachers. Vice versa, professional development of teachers without any guarantee of the institutional conditions for putting a real pedagogical renewal into practice would end up by hindering any large-scale diffusion of such a renewal.

To act on both these conditions we must be aware of the rapid changes in everyday learning life and consequently of how the figure of the 'e-teacher' should be 'modelled' to produce someone who is able to use NMTs both for their students' learning and for their own continuous professional development.

An example of the use of network technologies for teachers' continuous professional development is the central focus of Chapter 7, where Manuela Repetto presents the ALADIN project, discussing the informal approach adopted (based on action research), the results achieved and the critical factors which would help or hinder such an initiative in becoming a continuous education practice in the professional development of teachers.

In the final chapter (Chapter 8), Cathleen A. Norris and Elliot Soloway make some conclusive reflections on the central theme of the book, focusing on the concept that NMTs might be the ideal tool for true pedagogical innovation based on learning-by-doing. The authors look to the future and declare their conviction that with NMTs the opportunity to change education is literally 'at hand'.

We wish to thank all the authors who have contributed their precious scientific knowledge and experience to the writing of this book, which is addressed to all those who like us desire to constantly increase their understanding of the best educational use of NMTs for integrating formal and informal learning.

Guglielmo Trentin and Manuela Repetto

Bibliography

Norris, C. and Soloway, E. (2011) 'Learning and schooling in the age of mobilism'. *Educational Technology* 51(6): 3–10.

About the contributors

Ben Bachmair was until 2008, Professor of Pedagogy, Media Education and Media Enhanced Learning at the University of Kassel. Currently he is a visiting professor at the Institute of Education, University of London. He is a founding member of the London Mobile Learning Group. His specialisms include mass communication and education, media and learning, mobile learning, media socialisation, media reception and the role of media in cultural development.

Debra Bateman is an experienced educator across a range of contexts. She is especially interested in futures studies, how educational practices are shaped and how policies are developed and enacted. In teaching and learning, Debra is particularly interested in notions of innovation and playful pedagogies. Her creative use of wide-ranging technologies and innovative learning environments was recognised in 2010 when she was awarded an ALTC Citation for Outstanding Innovation in Teaching and Learning. In 2011, she received two National Awards for Teaching Excellence.

Roger Bateman is a senior lecturer in the Faculty of Arts, Computing, Science and Engineering at Sheffield Hallam University, where he teaches undergraduate and postgraduate product design. He is also works within The Culture, Communication and Computing Research Institute and the university's Art and Design Research Centre. From 2004 to

2010 he worked in the School of Design and Visual Arts at Unitec. Roger's research interests include mobile learning, Web 2.0, sustainable product design and design for health.

Thomas Cochrane is an academic adviser and senior lecturer in educational technology at the Auckland University of Technology, Centre for Learning and Teaching. From 2004 to 2011 he worked at Unitec as an academic adviser in the field of e-learning and learning technologies. His research interests include mobile learning, Web 2.0 and communities of practice. Thomas has managed and implemented over 35 mobile learning projects, with a recent focus upon Android and iOS smartphones and the iPad as catalysts to enable student-generated content and student-generated learning contexts, bridging formal and informal learning environments.

AnneMarie Conley is an assistant professor of education at the University of California, Irvine. She studies how students are motivated to learn, particularly in the context of science and maths education. Her California Motivation Project investigates how the will to learn develops during adolescence, and how this motivation influences how much students learn and achieve. This research, funded by several grants from the National Science Foundation, extends to assessing teachers' motivation to learn in professional development contexts and the influence of teachers' motivation on student achievement.

John Cook is Professor of Education at the University of the West of England. Previously he was Professor of Technology Enhanced Learning and Director of the Learning Technology Research Institute at London Metropolitan University. He has a specific interest in the related areas of informal learning, mobile learning in all sectors, augmented contexts for development, user-generated contexts and work-based learning. John is a founding member of The London Mobile Learning

Group. He was Chair/President of the Association for Learning Technology (2004–06) and Chair of the Association for Learning Technology's Research Committee (2008–12).

Ola Erstad is Professor of the Department for Educational Research, University of Oslo. He has been working within the fields of media and educational research with a particular focus on media literacy and digital competence. He leads the research group TransAction-Learning, Knowing and Identity in the Information Society. He is involved in several research networks on digital media and education nationally, across the Nordic countries in Europe and internationally.

Cathleen A. Norris is a Regents Professor in the College of Information, Department of Learning Technologies at the University of North Texas. Her research agenda of helping teachers to employ mobile technologies more effectively in the twenty-first-century classroom has been shaped by 14 years in K-12 classrooms. Cathie has been President of the International Society for Technology in Education, and the President of the National Educational Computing Association. Cathie is co-founder of GoKnow, Inc., a Dallas-based company that supports K-12 in using mobile learning devices.

Norbert Pachler is Professor of Education and Director: International Teacher Education at the Institute of Education, University of London. In 2007 he founded the international, interdisciplinary London Mobile Learning Group, which he convenes. The group comprises researchers in the fields of cultural and media studies, sociology, (social) semiotics, pedagogy, educational technology, work-based learning and learning design. Norbert is widely published and his research interests include the application of new technologies in teaching and learning, teacher education and development and all aspects of foreign language teaching and learning.

Manuela Repetto is a researcher at the Institute for Educational Technology of the Italian National Research Council. She has been involved in a number of projects in the field of e-learning, teacher training, lifelong learning and special education. Her research interests include the definition of models and methods for studying virtual learning communities, and the use of web-based approaches to innovate learning and professional practices.

Solveig Roth is a PhD student at the Department of Educational Research, University of Oslo. Her work aims to theorise how young Norwegian girls learn, negotiate and reflect upon knowledge within and across a variety of social contexts in their everyday lives, and how this affects their learning identity and future orientation. She has also worked as a secondary school teacher, in teacher education at university college level and as a clinical pedagogue.

Elliot Soloway is Arthur F. Thurnau Professor in the Department of Computer Science and Engineering, College of Engineering, School of Education and School of Information, University of Michigan. His research is guided by the vision of achieving universal 1:1 in schools across the globe through low-cost handheld mobile and networked devices. He was voted Outstanding Teacher of the Year by University of Michigan undergraduates in 2001, and presented with the Distinguished Teacher of the Year award by the EECS College of Engineering HKN Honor Society in 2004 and 2011. Elliot is a co-founder of GoKnow, Inc. He is currently Chair and Grand Poobah of ISTE's Special Interest Group on Mobile Learning.

Cathy Tran is a PhD student in the learning, cognition and development programme at the University of California, Irvine. She studies how motivation and cognitive theories apply to the design of educational media. She leads an interdisciplinary team of faculty and students in applying research theories to the design of an educational mobile app. Prior to graduate school, she worked in research, development and evaluation for children's television, museum and educational software projects.

Guglielmo Trentin is a senior researcher and project manager at the Italian National Research Council's Institute for Educational Technology. His studies have largely focused on the use of network technology in formal and informal learning. In this field he has managed several projects and scientific activities, developing technological applications and methodological approaches to support networked collaborative learning. He teaches network technology, knowledge flow and human resources development at the University of Turin. He is presently the principal investigator of Wiring Individualised Special Education (WISE), a strategic research project funded by the Italian Ministry of Universities and Research that studies the use of network and mobile technology in homebound special education.

Mark Warschauer is Professor of Education and Informatics at the University of California, Irvine and Director of the university's PhD in education programme. He has previously taught or conducted research in Russia, the Czech Republic, Egypt and Japan. He is the founding editor of *Language Learning & Technology* journal. His most recent book is *Learning in the Cloud: How (and Why) to Transform Schools with Digital Media*. He is currently investigating the use of digital tools to promote improved literacy and learning among diverse K-12 students.

Julie Willems holds qualifications in nursing, the humanities and education. She has worked in all education sectors in Australia including secondary, community, technical and further education, and higher education, as well as in face-to-face, virtual, flexible and online contexts in the latter. Her research examines equity, access and diversity in various contexts of education, including distance, flexible, mobile, virtual and e-learning, in addition to the technologies of Web 2.0.

The editors may be contacted at:

Guglielmo Trentin
Institute for Educational Technology
National Research Council
Via De Marini, 6 – 16149
Genoa
Italy

E-mail: *trentin@itd.cnr.it*

Tapping the motivational potential of mobile handhelds: defining the research agenda

Cathy Tran, Mark Warschauer and AnneMarie M. Conley

Abstract: As mobile technology becomes increasingly prevalent, educators are looking for ways to tap its educational potential. This chapter argues that one of the biggest strengths of handheld mobile devices is their potential to increase motivation for learning. Elements such as their small size, geolocation awareness, instant information access and tactile features contribute to a portable and personal experience that has the potential to motivate learners. The chapter makes the link between motivation and technology by first reviewing existing literature on the motivational implications of educational technology in general and then discussing how the unique features of mobile handhelds may further increase motivation for learning. In defining handhelds, we exclude laptops and refer to smaller devices such as smartphones and tablets. Keeping with the theme of this volume, the chapter then considers both formal and informal environments as it discusses a research agenda and methodological considerations for empirically exploring the relationship between mobile use and learner motivation.

Keywords: motivation, handheld technology, self-determination theory, embodied cognition, situated learning

Introduction

This chapter sets forth a proposition that there is promising untapped potential for applying motivation theories to learning with handhelds. Part of this optimism stems from the growing literature on how games motivate goal-directed behaviour (see discussion in Gee, 2003; Przybylski et al., 2010). Every day, about 30 million players of the mobile app *Angry Birds* invest 5 million hours in launching birds at towers of bricks and rocks in an effort to destroy pigs (Rovio, 2011), and more than 30 million players harvest their crops in the social network game *Farmville* (Cashmore, 2010). When people are asked why they sacrifice other leisure, and perhaps non-leisure, activities to engage in such games, the immediate and most obvious response is simply because they are fun. Unpacking what makes games fun can help in the design of mobile technology that fosters highly-motivated learning.

The increasing prevalence of mobile technology is evident in the statistics: among 8 to 18-year-olds between 2004 and 2009, cell-phone ownership increased from 39 per cent to 66 per cent (Rideout et al., 2010). Nine out of ten 18 to 29-year-olds in the USA own a cell phone, and among those owners, 65 per cent access the internet on their mobile devices (Smith, 2010). Cell phones have evolved to be more than just a tool to talk, as young people now spend more time listening to music, watching television and playing games on their cell phones than they spend talking on them (Rideout et al., 2010). With the surge in ownership of tablets, educators are, more than ever, looking for ways to integrate mobile devices in the design of learning environments. What is missing from this integration, however, is a learning theory that is specific to mobile technologies and will allow for the effective design of educational environments with mobile handhelds. The potential for mobile handhelds to contribute to lifelong, contextualised learning rests on the establishment of principles that highlight and exploit the distinctive features of those devices, yet the formation of these principles is still in its infancy (Shuler, 2009).

A theoretical model of motivation and mobiles can draw heavily on psychological theories of motivation. While there are clear developmental considerations, this chapter focuses on components and processes that are relevant across the lifespan. In defining mobile handhelds, we exclude laptops and refer to devices that can be used on the go without sitting down, such as smartphones and tablets. Their small size, cordless nature, geolocation awareness, instant information access and tactile features (e.g. touchscreens, vibrations) contribute to a portable and personal experience that has the potential to motivate learners and allow them to build their knowledge across formal and informal settings. Motivation is a worthwhile construct to target as some scholars argue that motivational factors play a larger role than academic performance in predicting continued learning. For instance, motivation for learning psychology was more predictive of subsequent course taking and majoring in psychology over a seven-year span than were grades from an introductory college psychology course (Harackiewicz et al., 2002). Similar patterns have been found for middle school and high school students (Harackiewicz et al., 2002; Hidi, 1990; Hidi and Harackiewicz, 2000; Hidi and Renninger, 2006). Although research on motivational theories and their applications to classroom education has generated thousands of journal articles, there is limited empirical evidence about whether these theories apply equally well in the context of learning with handhelds.

Before delving further into the discussion of the integration of motivational theories and learning with handhelds, we first need to clarify what we mean by the term 'motivation'. Motivation is 'the process whereby goal-directed activity is instigated and sustained' (Pintrich and Schunk, 2002: 5). Put simply, it is what influences individuals to take action and keep at it when the going gets rough. Motivation is a multifaceted construct as there are many factors that move people; for instance, it may be personal interest, a bribe, or a desire to impress others. As such, motivation can be internal (intrinsic motivation) or it can be externally pressured (extrinsic motivation). Compared with those pursuing an activity for external rewards like money, research has shown that intrinsically motivated individuals act more creatively, enjoy the activity more, and process information more carefully and completely (Ryan and Deci, 2000). Accordingly, this chapter focuses on how mobile handhelds can increase internal or intrinsic motivation and adopts the perspective of self-determination theory (SDT), which posits that intrinsic motivation arises when learners' needs for competence, autonomy and relatedness are met (Deci and Ryan, 2000; Ryan and Deci, 2000). That is, to be motivated, people need to feel that they are: (1) capable of understanding the presented material, (2) in control of their environment, and (3) socially connected. While we highlight SDT, other motivational theories have also been applied to the design of educational technology. For example, expectancy-value theory (e.g. Eccles, 1987, 1993; Eccles et al., 1989) influenced the design of a mathematical game to encourage students to pursue careers in science, technology, engineering and mathematics (Chen et al., in press; Tran et al., 2012). Achievement goal theory (Pintrich et al., 2003) was used to understand the role of achievement badges in a robotics program (Abramovich et al., 2011). A review article by Pintrich (2003) details these motivational theories and other prominent ones that have potential applications to the design of educational technology.

SDT is particularly applicable to the study of learning with mobile handhelds because their use is often driven by intrinsic motivation as people typically engage with them by choice. Handhelds include portable and personal features that are well suited to address the need for autonomy, competence and relatedness. Portable features stem from the cordless nature and small size of cell phones as they fit snugly inside pockets, allowing for the digitally-tethered life. This constant access eliminates the constraint of being able to access data only at a certain location such as a library or desktop computer, while cloud computing allows for personal data to transfer seamlessly across desktops and mobile handhelds. While desktop computers have advantages in some situations, such as when a group explores a large dataset together on a screen, mobile handhelds are particularly suitable for informal environments and learning on the go. Features like geolocation awareness take advantage of the portable nature by embedding geotags to alert users of potential details of interest in their geographical vicinity.

Personal features that allow mobile handhelds to be responsive by touch and voice have been enhanced by recent technological developments. Some devices react to shaking, rotating and tilting. In addition, advanced voice features are contributing to the role of handhelds as personal assistants. These portable and personal features can satisfy the three SDT needs to contribute to inherently enjoyable activities that are therefore intrinsically motivating. Using handheld devices to address the psychological needs of competence, autonomy and relatedness can foster sustained engagement and positive learning outcomes for both children and adults. We discuss below how a motivation theory maps onto educational technology and then propose an agenda for investigating handhelds and motivation for learning by discussing future research questions and methodological considerations for empirically exploring the relation between mobile use and learner motivation in both formal and informal environments.

Educational technology and motivation: past, present and future

Rapid advances in technology have historically brought with them new tools to increase learner motivation. When educational videos entered the market, at-home learners had autonomy over the pace of instruction as they could pause, rewind or fast-forward a tape. With the advent of personal computers, software provided programs that allowed for rapid feedback and individualised instruction in a classroom of one teacher to dozens of students. In the 1990s, as the internet became more mainstream, learners experienced autonomy through the choice of sequence with the multitude of hyperlinks that connect online information in a nonlinear fashion. The internet addressed individuals' needs for relatedness by providing an audience for creations ranging from blog posts about daily life to YouTube clips showcasing singing talents. Recent digital media platforms have enhanced this trend of online sharing through the remix mash-up culture of fusing clips of others' work in the creation of new personal media.

Nowadays, at the forefront of new technology is the rapid adoption and growing sophistication of mobile handhelds, which provide a new platform for designing learning environments. Traditional computer use in educational settings usually requires a trade-off. Taking advantage of the benefits of workplace technology typically requires that students work quietly and independently in their chairs, devoid of most social interactions. Mobile handhelds, however, can potentially allow for the best of both worlds as learning environments can be designed to harness the motivational affordances of both in-person and technology-enhanced experiences.

Through a review of existing literature, we identify several design elements in educational technology that have been successful in attending to the motivational needs among learners. We then discuss how mobile handhelds are uniquely able to contribute to those needs above and beyond what previous technology has done. In doing so, we suggest ways of taking advantage of the affordances of mobile handhelds to design learning environments in ways that allow learners to feel like they can successfully learn, have control and be connected with others.

Competence: rapid feedback and adaptive instruction

The need for competence refers to the desire to experience oneself as capable of developing key skills and abilities. This desire leads people to seek challenges that are optimal for their capacities and to persistently attempt to maintain and enhance those skills and capacities through activity. Rather than being an attained skill or capability, competence is a felt sense of confidence (Ryan and Deci, 2002). For students to feel competent, for instance, they need to believe that they have the skills and resources to meet the challenges in their learning environment. Research shows that optimal challenges, feedback and freedom from demeaning evaluations are predictors of increased sense of competence and thus greater intrinsic motivation (Ryan and Deci, 2000).

Educational technology has addressed the need for competence through the provision of rapid feedback and adaptive instruction. Technology often allows learners to receive feedback more rapidly than they would in a traditional classroom by instantly informing students of the correctness of their answer and providing information as to why such a response is correct or incorrect. Feedback is critical for effective learning because it allows students to monitor their learning and actively evaluate their strategies and their current level of understanding (Bransford, 2000). Technology has also allowed for more efficient adaptive instruction, structuring the learning experience such that challenges gradually increase or decrease in line with students' progress through the programme. This balance is critical to learning; if the tasks are underwhelming, they will lead to boredom, and if they are overwhelming, they will lead to frustration. Vygotsky (1978) posited that students learn most when they are continually challenged within a zone of proximal development, in which the skills and knowledge required for successful completion of the task are slightly out of reach of where students are presently functioning but can be successfully completed through scaffolded assistance (for a review of educational technology and scaffolding, see Sharma and Hannafin, 2007).

Mobile handhelds provide unique affordances for providing feedback and adaptive instruction. Feedback can be provided to align with the learners' specific location. In the mobile app *Star Walk*, for example, the program instructs users to tilt the camera left, right, up or down to track and discover thousands of stars, planets and constellations. The device's built-in location detection determines the user's location, while its compass and accelerometers track the tilts of the screen as it displays the night sky in real time. In addition, adaptive instruction can uniquely be provided on mobile handhelds by building in constraints of how users can manipulate digital objects. For example, if a learner struggles with a spatial task because he or she cannot rotate an object inside his or her head, the program can allow for objects to be freely rotated, and this option to rotate can be constrained as the learners perform better on the spatial tasks.

Autonomy: learner-paced and tactile features

Feelings of competence alone will not enhance intrinsic motivation unless learners also feel that they have some control over their learning (Fisher, 1978; Ryan, 1982). When non-controlling teachers support this need for autonomy, students experience more intrinsic motivation, curiosity and desire for challenge (Deci et al., 1981; Flink et al., 1990; Reeve et al., 2004; Ryan and Grolnick, 1986). Educational technology, through purposeful design, can provide affordances that allow students to control several aspects of their environment more easily, enhancing both autonomy and learning. People learn better when a multimedia message is presented in user-paced segments rather than as a continuous unit, particularly when the material is complex, the lesson is fast-paced, and the learner is inexperienced in the domain (Mayer, 2002). Educational software and the internet enable nonlinear access to multiple forms of information, allowing students to navigate more freely among content, define their desired sequence of information uptake, and explore information in more depth on demand. A meta-analysis on this nonlinear form of hypermedia and learning found that the sequence controls afforded by hypermedia positively affect learning outcomes (Dillon and Gabbard, 1998).

Mobile handhelds provide unique affordances for providing learner-paced and tactile features to enhance autonomy. For example, in *iTacitus*, panning the mobile device with the app at a historical ruin places three-dimensional objects such as statues into the scene, overlaying the real scene on the mobile screen. This landscape is annotated, and learners can choose when to show more informational texts, images or videos about a particular spot or object as well as how much information to show. Tactile features of mobile handhelds also play a role in addressing learners' need for autonomy. Mobile handhelds allow for more natural processes of digital construction in that learners can drag and drop with their fingers rather than use a mouse. In the mobile app *Monster Physics*, for instance, users are in control as they build their own car, crane or rocket ship and then tinker and tweak their constructions so that they move as desired. This experience maps onto the notion that 'learning is understood as a constructive process of conceptual growth, often involving reorganization of concepts in the learner's understanding, and growth in general cognitive abilities such as problem-solving strategies and metacognitive processes' (Greeno et al., 1996: 16). There is an assumption that understanding is gained by an active process of construction rather than by passive assimilation of information. With respect to handhelds, autonomy may be enhanced by the touchscreen because the way that users move and manipulate objects on a handheld, compared with on a desktop, are more analogous to real life, which may give them a sense that they are in direct control. Mobile devices are also more personal in that they automatically save and open up to the user's last screen such that the stop-and-start interaction with handhelds is more seamless, allowing for short bursts of usage. The user is more autonomous in choosing how and when to interact with the handheld as the result of the ease of starting and closing applications as well as its portability.

Relatedness: audience and collaborative learning

A sense of relatedness and belongingness, such as students feeling respected and cared for by the teacher and peers, facilitates intrinsic motivation (Ryan and Deci, 2000). The internet provides new opportunities for social connectedness by allowing for a greater audience with whom students can share information and by providing a structure for computer-supported collaborative learning (e.g. Warschauer, 1996, 1997). One domain for which this affordance is particularly applicable is writing. The benefits of writing to an audience have a long history in the research literature. Redd-Boyd and Slater (1989), for example, conducted an experiment in which undergraduate students were either assigned a specific reader or not assigned a reader for a persuasive essay. Their findings suggest that assigning an audience increased students' interest and effort in writing the essay. Technological affordances such as blogging and writing fan fiction now open more avenues for learners to share their knowledge with an audience (see review in Warschauer and Grimes, 2007). Research has also shown that cooperative teams think more deeply and retain information longer than students who work individually (Johnson and Johnson, 1986). Technology has allowed for more opportunities for collaborative learning and thus more opportunities to address the need for relatedness.

The portable nature of handhelds makes it easy for learners to simultaneously play on the same screen, supporting learners' needs for relatedness. In the mobile app *Toontastic*, for example, drawing tools allow children to create and animate cartoons. A touchscreen tablet platform makes it easier for users to collaborate because switching back and forth does not requiring passing the mouse. These cartoons can be developed individually and shared with an audience or be developed in conjunction with family and friends, as different users can animate and be the voice of different characters. Users can then share their cartoons on the web and watch and comment on the other cartoons, increasing a sense of relatedness.

Although we have independently discussed competence, autonomy and relatedness, we emphasise that the design of learning environments should aim to address all three. Intrinsic motivation requires that the needs for competence and autonomy are both met; relatedness further enhances that motivational state (Ryan and Deci, 2000). Supports for competence will enhance intrinsic motivation only if administered in a way that is autonomy supportive (Ryan, 1982). Similarly, supports for relatedness will enhance intrinsic motivation only if the involved others are autonomy supportive (Grolnick and Ryan, 1989). Successful integrations of SDT components are illustrated in the *Live Long and Prosper* handheld game that teaches students about genetics (Klopfer, 2008). To live as long as possible and to reproduce as much as possible, players must determine how specific traits are influenced by their genes and gain an understanding of dominant and recessive genes. To do this, players trek around the classroom to select other students with whom to mate via their handhelds in an effort to figure out their genome. When players 'die', the teacher moderates a discussion to collect hypotheses about the cause of death before the next round of reproduction starts. This example shows how handhelds allow students to be in control as they learn with mobile technology through face-to-face social interaction, addressing the needs of autonomy, social relatedness and competence.

Proposed research strands for motivation and mobile handheld technology

Research on what makes learning with handhelds motivating has been limited, in part, because motivation has been conceptualised as distinct constructs or as distinct design components. There has been little deliberate incorporation of motivation theory, which views motivation as composed of multiple complex and interacting components. Some mobile games designed to increase motivation have done so by incorporating individual components such as rules, goals and feedback, but have paid insufficient attention to the way these features interact. While these components have the potential to be motivationally beneficial, understanding the motivational theory through which they operate is essential to creating learning environments that support learners' motivation.

In linking SDT with learning with handhelds, we draw on literature in cognitive (e.g. embodied cognition; for a discussion see Wilson, 2002) and social psychology (e.g. situated learning; for a discussion see Brown et al., 1989). In the section below, we identify three strands of research that explore how the portable and personal features of mobile handhelds can impact the design of learning environments. For each strand, we describe the motivational issues at hand and propose a series of research questions that need to be addressed to advance the field. While there are many other possible research directions examining motivation and mobile learning, we find these ones to be the most pressing because they link the affordances of mobile devices to robust, theory-driven literature on what is known about learning.

The first research strand, bringing the computer to the physical context, highlights the importance of learning in the real world and the ways in which mobile handhelds can facilitate this learning. The second strand explores how simultaneous in-person interaction and technology use can combine to potentially enhance motivational outcomes. The third strand describes how tactile features of mobile devices can be leveraged to address the importance of physical movement and feedback in motivation and learning.

Bringing the computer to the context

Whereas desktop computers help contextualise learning by bringing real-world context onto the screen, mobile learning brings the screen into the physical context via augmented reality. Interactions with desktops lack the physicality, concreteness and connection with real-world environments made possible with the use of mobiles. The portability of mobile technologies makes it possible for new and enhanced learning opportunities that take place outside of school. From a motivational stance, bringing the computer to the real world means that the target learning environments can be places in which people already choose to engage, allowing for an increased sense of autonomy. Furthermore, people choose how much supplementary information they want and when they want it. For example, a traveller may pass a historical site and wonder what it looked like when it was in use and whether anything significant happened there. An augmented reality mobile app can place historical objects into the scene, overlaying the current scene that is shown on the mobile screen. This landscape can be annotated, allowing users to tap on parts of the images for which they would like more information. The historical ambiance of the location can be further recreated with audio overlays, adding multiple dimensions to the sense of reality. This type of connection to reality makes it easier for students to identify with the material and makes learning more interesting and meaningful, which can increase motivation via deep thinking and increased competence.

While there is much to like about bringing the computer to the context, there are also potential drawbacks. Handhelds can allow for numerous options to enhance the understanding of outside environments, but too much choice of supplementary information within those environments could result in decreased competence (e.g. Iyengar and Lepper, 2000). Do learners know how to pace their consumption of knowledge appropriately? Is the provision of free choice worth giving up some of the cognitive advantages of structured sequence to scaffold experiences for learners? Would too many choices overwhelm learners? This is of relevance to mobile learning in particular as handhelds open learners to a wide array of choices to supplement their experience in the world. To harness the motivational potential of the new choices that mobile handhelds offer, there is a need to explore the degrees of choice that are optimal for different types of individuals and different situations. Research on choice and mobile technology, or on choice in informal learning environments in general, is limited.

Contextual learning on a desktop computer has its motivational benefits (e.g. case study in Dede, 2009), and to our knowledge, no research has compared the advantages of contextual learning on a desktop with in-situ learning using a mobile device. Is bringing the computer to the real world more motivational than bringing real-world context to the computer? This may be the case, as it cannot be assumed that the learning environment on a desktop computer that is designed to reflect reality is actually experienced by students as being authentic (Gulikers et al., 2005). Bringing the computer directly into the physical world takes this guesswork out of the equation. On the other hand, there are situations in which a desktop computer may be more motivational, such as in a graphic-intensive interactive educational narrative, as is the case for many multi-user virtual environments. Future empirical work can inform the types of situations in which the unique features of mobile handhelds are most advantageous for motivated learning.

Engaging in simultaneous in-person interactions and technology use

Because mobile technology is not tethered to a desktop computer it can be brought into a space where students can interact in person while interacting individually with technology. In traditional desktop computer settings, there is often a trade-off between in-person interaction and the individualised cognitive scaffolding that technology offers. Motivation theory posits that combining in-person interactions and technology through mobile handhelds would be intrinsically motivating because students can receive cognitive scaffolds through technology as they interact with each other. In-person interaction addresses the need for relatedness, especially when the social connectedness in the learning environment is critical to advancing other participants' knowledge (Ryan and Deci, 2000). In the genetics game previously mentioned, students are not able to progress as individuals – they need to mate with one another through handhelds to understand their own genomes and to reproduce; as such, the social interaction acts as a core component of the activity rather than as a supplementary component. The feedback provided by this interactive nature of the learning experience addresses the need for competence and guides the students' thinking and subsequent mating moves, enhancing a sense of autonomy.

Understanding the motivational benefits of combining in-person interactions and technology use requires the exploration of numerous unknowns. For instance, does combining the two result in additive benefits? Or is it an exponential benefit? Or is it perhaps less than the sum of their parts? Although no work, to our knowledge, has answered these questions, a meta-analysis of research about online learning published from 1996 to 2008 suggests that when curriculum includes both online and in-person instruction (though not at the same time), students have a larger learning advantage relative to purely in-person instruction or purely online instruction (Means et al., 2009). Mobile learning adds a new dimension to this as the blending of in-person interaction and technology use can be simultaneous. The effects of this simultaneous interaction may vary among individuals. For individuals who have a higher need for relatedness, for instance, having more social connectedness through in-person feedback may be more beneficial than attaining quicker feedback through the technology only. For novice learners, the cognitive scaffolding afforded by the technology may be extremely important, and at that stage, having an element of simultaneous social interaction may be a distracting feature.

A second area of future work is exploring if there is an advantage to combining in-person interactions with technology use (which a mobile environment facilitates) over virtual interaction such as via webcam. In-person interactions are not impossible on desktops as two people can sit side by side, but it may occur more naturally using mobile devices. Although we were unable to find research that addressed such a comparison, previous work has provided insights about the effects of each model individually. A robust body of literature supports that in-person social interaction provides more prompts and inputs for cognition (see review in Webb, 1989), addressing the need for competence. More recent work in the virtual environment suggests that merely adopting a social frame of mind affects learning. Okita et al. (2008) conducted an experiment in which participants had identical interactions with a virtual-reality avatar that was represented as being human-controlled or computer-controlled. Compared with those who thought they were interacting with the computer, participants who thought they were interacting with a human exhibited more arousal and attention as well as higher scores on a post-test of science concepts. The perceived level of socialness and not just the interactions themselves appears to matter.

We expect that future research in this area will suggest that in-person interaction is linked with greater motivational benefits than virtual interaction because it is more fast-paced and flexible. In addition, it allows for the quick interpretation of gestures, facial expressions and other audio-visual clues as well as quick and easy reference to drawings, charts or physical artefacts. This rich human and physical environment allows learners to follow better what someone is saying and allows a teacher or mentor to see the interactions and more quickly diagnose how students are following an activity and thus make rapid adjustments to get a point across better. While all these types of interactions can be simulated in various ways via desktop computer, those computer-based interactions generally take much more time and are not nearly as rich in communicative content (Warschauer, 2003). As such, the rich features of in-person communication coupled with technology scaffolds that handhelds afford may increase motivation by addressing the needs for competence and relatedness.

Leveraging tactile features of mobile handhelds

Personal tactile features of handhelds, such as response to touch and tilts, allow for technology-enhanced learning environments that are physically interactive. Kinaesthetic modes of representation and expression are an important dimension of learning because, as the embodied cognition literature posits, knowledge is rooted in our physical interactions with the world and not just in the brain (Wilson, 2002). For example, Alibali and Nathan (2011) found that gestures that show relatedness between mathematical concepts were predictive of teachers' and students' mathematical understanding. Ramani and Siegler (2008) showed that preschool students gained understanding of numerical magnitude by playing board games that allowed them to interact physically with a number line by moving tokens along it. The authors posited that the tangible nature of the board game might have helped children develop a mental number line.

Educational designers have begun to leverage the capability of mobile technologies to create kinaesthetic learning experiences. For example, a display can be pinched to zoom in or out on an image and can detect when it is being moved or tilted. The mobile game *Motion Math HD*, which teaches students about fraction estimations, has taken advantage of the tactile feedback mechanisms. In the game, learners tilt their phone or tablet to guide a bouncing ball labelled with a fraction on to the appropriate spot on a number line. Once the fraction-ball lands at approximately the right position, a new fraction appears on the bouncing ball, and the student must tilt the device to bounce the ball to a new location. Increasingly embedded into mobile devices, accelerometers offer a unique and intuitive interaction style. Although a more detailed review of the embodied cognition literature is out of the scope of our chapter, Denham et al. (2012) provide more details on how theories of embodied cognition can inform the design of mobile learning environments.

Future research on mobile technology can explore how physical manipulation through digital handhelds contributes to motivational needs. For example, does digitally manipulating objects on a touchscreen offer the same motivational benefit as physically manipulating objects in real life? Manipulating representations can support task performance by offloading cognitive processes to the perceptual system (Wilson, 2002). As such, manipulations can address the need for competence by providing assistance in learning conceptual and spatial knowledge. As digital manipulations still allow learners to use their hands to construct different visual representational states, we predict that the motivational outcomes for digital manipulations will be very close to the benefit of the manipulation of physical objects. Furthermore, what do digital manipulations on handhelds add to learning, above and beyond physical manipulations? Unlike physical manipulations, digital manipulatives can be linked with feedback and a record of prior actions. Technology also provides the means to constrain actions on handheld screens if and when this is beneficial for learning. For example, games can be designed to foster particular strategies by changing the relative ease of moving objects in particular ways. Manches et al. (2010) demonstrated how constraining actions using a digital interface such that children could move only one object at a time resulted in the use of more efficient strategies in solving a mathematical problem compared with using physical representations. While it is possible to constrain actions using physical devices, it is typically more difficult to design ways to vary such constraints. In addition, handhelds can support problem-solving by introducing a delay after each action to encourage reflection. Therefore, the learner is still autonomous but is within a structured environment, and these affordances may better address the need for competence. These research issues and related questions are summarised in Table 1.1.

Table 1.1 **Research agenda for motivation and mobile handhelds**

Research issues	Research questions
Bringing the computer to the context	Is bringing the computer to the real world more motivational than bringing real-world context to the computer? Is the provision of free choice worth giving up some of the cognitive advantages of structured sequence to scaffold experiences for learners? Do learners know how to pace their consumption of knowledge appropriately? Would too many choices overwhelm learners?
Engaging in simultaneous in-person interactions and technology use	Does combining in-person interactions and technology use result in additive benefits? Or is it an exponential benefit? Or is it perhaps less than the sum of their parts? Do in-person and virtual interactions address the motivational need for relatedness differently? What kinds of conversations do learners carry out in in-person vs virtual environments?
Leveraging tactile features of mobile handhelds	Does digitally manipulating objects on a touchscreen offer the same motivational and cognitive benefits as physically manipulating objects in real life? Are there motivational advantages of using fingers on a touchscreen over using peripherals such as a mouse on a desktop? What do digital manipulations on handhelds add to motivation and learning, above and beyond physical manipulations?

Methodological considerations for research in mobile learning and motivation

To advance theories of mobile learning, we need to advance research methodologies specifically aligned to the unique attributes of mobile learning. The evaluation of mobile learning is unusually challenging because evaluation techniques are based largely around formal, sedentary and traditional learning (Traxler, 2007). There are inherent difficulties in following learners with mobile devices as they are constantly on the move. Embedding assessments within the technology itself provides a solution. Researchers and educational designers of mobile learning can take advantage of the opportunity for embedded assessment through the use of methods like experience sampling, data logging and affective computing.

Experience sampling, which allows researchers to collect systematic data about what a person does, thinks and feels during daily life, can allow for continuous longitudinal data collection in mobile learning research. This method measures participants' location, activity, and affective and cognitive experiences at pre-set intervals (Larson and Csikszentmihalyi, 1983). It is particularly valuable for eliciting the subjective experiences of persons interacting in their natural environments. The record of daily experiences produced is similar to that recorded in time diaries, but differs in that respondents record their experiences immediately upon having them. These immediate responses reduce the error created by the reconstruction of experiences when recorded well after their occurrence (Larson and Csikszentmihalyi, 1983) and have been demonstrated as both reliable and valid (Csikszentmihalyi and Larson, 1992).

Traditionally, participants carry a pager over the course of several days or weeks, and the pager beeps several times a day to elicit a report about their experience. With the omnipresence and popularity of cell phones, it is possible to gather data over a longer period of time because participants are not required to carry an additional device. Experience sampling can provide insight into learners' motivational states by eliciting information about the learners' level of autonomy, competence and relatedness throughout the day. Furthermore, if the mobile device itself is being studied as a learning tool, the solicitation of learners' experience can be timed with specific actions that occur on the handheld to assess the motivational state at that time point. Researchers can also use geotagging to target specific real-world locations and elicit experience sampling data. This allows for real-time motivational data for mobile learning across context in both formal and informal environments (Muukkonen et al., 2008).

Mobile handhelds also allow for motivational assessments to be built in, creating a more seamless evaluation that is not disruptive to the learning environment. Automatic data logging, for instance, can be activated to collect data about user activity with the different mobile applications on the handheld. Automated logging can record cumulative application usage time and the relative distribution of run time of each application (Trinder et al., 2009). These log data can then be triangulated with additional assessments to provide insight about user engagement and motivational states.

Furthermore, tools from the affective computing field can be embedded in mobile devices for motivation assessment. Recently developed computational models enable real-time analysis of cognitive-affective mental states from facial expressions (McDuff et al., 2011). Adaptive motivation is characterised by higher positive affect and lower anxiety (Pekrun et al., 2002), and these emotional states are expressed through facial expressions, gestures and other non-verbal channels. By streaming a continuous recording using cameras on mobile handhelds (with learners' permission), researchers can capture emotional states to gain insights about motivation without disrupting the learning experience.

Conclusion

While mobile handhelds have great potential for connecting contexts and increasing learner motivation, the strengths and limitations of mobile devices for learning are not yet well understood. The research agenda described in this chapter can help scholars and practitioners understand how best to strategically integrate mobile handhelds into the design of learning environments. To tap the motivational potential of handhelds, there is a need to know more about the implications of the unique affordances they provide: the ability to bring the computer to the physical context, simultaneous in-person interaction and technology use, and tactile features (see Table 1.1 for specific research questions). Mobile technology has become embedded in the fabric of everyday life and can be a valuable tool to bridge informal and formal learning environments.

Mobile handhelds have unique features that contribute to their intrinsically motivating nature, and there is great potential in taking advantage of these features to design more motivating learning environments.

Acknowledgments

The authors would like to thank Christine Capota, Arena Chang, Chris Dede, Stephanie Reich, Teomara Rutherford, Katerina Schenke and Adam Sheppard for their feedback on earlier drafts of this chapter.

References

Abramovich, S. J., Schunn, C. D., Higashi, R., Hunkele, T. and Shoop, R. (2011) 'Achievement systems' to boost achievement motivation', paper presented at Games+Learning+Society Conference 7. 0, Madison, WI, 15–17 June.

Alibali, M. W. and Nathan, M. J. (2011) 'Gestures in the classroom: What's the point?' in N. L. Stein and S. W. Raudenbush (eds), *Developmental Cognitive Science Goes to School*, New York: Routledge, pp. 219–34.

Bransford, J. (2000) *How People Learn: Brain, mind, Experience, and School*, Washington, DC: National Academies Press.

Brown, J. S., Collins, A. and Duguid, P. (1989) 'Situated cognition and the culture of learning', *Educational Researcher* 18(1): 32–42.

Cashmore, P. (2010) 'Farmville surpasses 80 million users', available at: *http://mashable.com/2010/02/20/farmville-80-million-users/* (accessed 8 November 2012).

Chen, J. A., Zap, N. and Dede, C. (in press) 'Using virtual environments to motivate students to pursue STEM careers: An expectancy-value model', in S. D'Agustino (ed.), *Immersive Environments, Augmented Realities, and Virtual Worlds: Assessing Future Trends in Education*, Hershey, PA: IGI Global.

Csikszentmihalyi, M. and Larson, R. (1992) 'Validity and reliability of the experience sampling method', in M. W. deVries (ed.), *The Experience of Psychopathology: Investigating Mental Disorders in their Natural Settings*, New York: Cambridge University Press, pp. 43–57.

Deci, E. L. and Ryan, R. M. (2000) 'The "what" and "why" of goal pursuits: Human needs and the self-determination of behavior', *Psychological Inquiry* 11(4): 227–68.

Deci, E. L., Nezlek, J. and Sheinman, L. (1981) 'Characteristics of the rewarder and intrinsic motivation of the rewardee', *Journal of Personality and Social Psychology* 40(1): 1–10.

Dede, C. (2009) 'Immersive interfaces for engagement and learning', *Science* 323(5910): 66–9.

Denham, A., Quick, J. M. and Atkinson, R. K. (2012) 'mLearning: An embodied perspective', *International Journal of Cyber Behavior, Psychology, and Learning* 2(3): 1–14.

Dillon, A. and Gabbard, R. (1998) 'Hypermedia as an educational technology: A review of the quantitative research literature on learner comprehension, control, and style', *Review of Educational Research* 68(3): 322–49.

Eccles, J. S. (1987) 'Gender roles and women's achievement-related decisions', *Psychology of Women Quarterly* 11(2): 135–72.

Eccles, J. S. (1993) 'School and family effects on the ontogeny of children's interests, self-perceptions, and activity choices', *Nebraska Symposium on Motivation* 40: 145–208.

Eccles, J. S., Wigfield, A., Flanagan, C. A., Miller, C., Reuman, D. A. and Yee, D. (1989) 'Self-concepts, domain values, and self-esteem: Relations and changes at early adolescence', *Journal of Personality* 57(2): 283–310.

Fisher, C. D. (1978) 'The effects of personal control, competence, and extrinsic reward systems on intrinsic motivation', *Organizational Behavior and Human Performance* 21(3): 273–88.

Flink, C., Boggiano, A. K. and Barrett, M. (1990) 'Controlling teaching strategies: Undermining children's self-determination and performance', *Journal of Personality and Social Psychology* 59(5): 916–24.

Gee, J. P. (2003) *What Video Games Have to Teach us about Learning and Literacy*, New York: Palgrave Macmillan.

Greeno, J. G., Collins, A. M. and Resnick, L. B. (1996) 'Cognition and learning', in D. C. Berliner and R. C. Calfee (eds), *Handbook of Educational Psychology*, New York: Macmillan, pp. 15–46.

Grolnick, W. S. and Ryan, R. M. (1989) 'Parent style associated with children's self-regulation and competence in school', *Journal of Educational Psychology* 81: 143–54.

Gulikers, J., Bastiaens, T. J. and Martens, R. L. (2005) 'The surplus value of an authentic learning environment', *Computers in Human Behavior* 21(3): 509–21.

Harackiewicz, J., Barron, K., Pintrich, P. R., Elliot, A. and Thrash, T. (2002) 'Revision of goal theory: Necessary and illuminating', *Journal of Educational Psychology* 94: 638–45.

Hidi, S. (1990) 'Interest and its contribution as a mental resource for learning', *Review of Educational Research*, 60(4): 549–71.

Hidi, S. and Harackiewicz, J. M. (2000) 'Motivating the academically unmotivated: A critical issue for the 21st century', *Review of Educational Research* 70(2): 151–79.
Hidi, S. and Renninger, K. A. (2006) 'The four-phase model of interest development', *Educational Psychologist* 41(2): 111–27.
Iyengar, S. S. and Lepper, M. R. (2000) 'When choice is demotivating: Can one desire too much of a good thing?' *Journal of Personality and Social Psychology* 79(6): 995–1006.
Johnson, R. T. and Johnson, D. W. (1986) 'Action research: Cooperative learning in the science classroom', *Science and Children* 24(2): 31–2.
Klopfer, E. (2008) *Augmented Learning: Research and Design of Mobile Educational Games*, Cambridge, MA: MIT Press.
Larson, R. and Csikszentmihalyi, M. (1983) 'The experience sampling method', *New Directions for Methodology of Social & Behavioral Science* 15: 41–56.
Manches, A., O'Malley, C. and Benford, S. (2010) 'The role of physical representations in solving number problems: A comparison of young children's use of physical and virtual materials', *Computers & Education* 54(3): 622–40.
Mayer, R. E. (2002) 'Multimedia learning', *Psychology of Learning and Motivation* 41: 85–139.
McDuff, D., Kaliouby, R. and Picard, R. (2011) 'Crowdsourced data collection of facial responses,' paper presented at 13th IEEE International Conference on Multimodal Interaction, Alicante, 14–18 November.
Means, B., Toyama, Y., Murphy, R., Bakia, M. and Jones, K. (2009) *Evaluation of Evidence-based Practices in Online Learning: A Meta-analysis and Review of Online Learning Studies*, Washington, DC: US Department of Education.
Muukkonen, H., Hakkarainen, K., Inkinen, M., Lonka, K. and Salmela-Aro, K. (2008) 'CASS-methods and tools for investigating higher education knowledge practices', in G. Kanselaar, V. Jonker, P. Kirschner and F. Prins (eds), *Proceedings of the 2008 International Conference for the Learning Sciences* (ICLS), Vol. 2, Utrecht: ICLS, pp. 107–15.
Okita, S. Y., Bailenson, J. and Schwartz, D. L. (2008) 'Mere belief of social action improves complex learning', in S. Barab, K. Hay and D. Hickey (eds), *Proceedings of the Eighth*

International Conference for the Learning Sciences, Vol. 2, Hillsdale, NJ: Lawrence Erlbaum Associates, pp. 132–9.

Pekrun, R., Goetz, T., Titz, W. and Perry, R. P. (2002) 'Academic emotions in students' self-regulated learning and achievement: A program of qualitative and quantitative research', *Educational Psychologist* 37(2): 91–105.

Pintrich, P. R. (2003) 'A motivational science perspective on the role of student motivation in learning and teaching contexts', *Journal of Educational Psychology* 95: 667–86.

Pintrich, P. R. and Schunk, D. H. (2002) *Motivation in Education: Theory, Research, and Applications*, Columbus, OH: Merrill.

Pintrich, P. R., Conley, A. M. M. and Kempler, T. M. (2003) 'Current issues in achievement goal theory and research', *International Journal of Educational Research* 39(4–5): 319–37.

Przybylski, A. K., Rigby, C. S. and Ryan, R. M. (2010) 'A motivational model of video game engagement', *Review of General Psychology* 14(2): 154–66.

Ramani, G. B. and Siegler, R. S. (2008) 'Promoting broad and stable improvements in low-income children's numerical knowledge through playing number board games', *Child Development* 79(2): 375–94.

Redd-Boyd, T. M. and Slater, W. H. (1989) 'The effects of audience specification on undergraduates' attitudes, strategies, and writing', *Research in the Teaching of English* 23(1): 77–108.

Reeve, J., Jang, H., Carrell, D., Jeon, S. and Barch, J. (2004) 'Enhancing students' engagement by increasing teachers' autonomy support', *Motivation and Emotion* 28(2): 147–69.

Rideout, V., Foehr, U. and Roberts, D. (2010) *Generation M2: Media in the Lives of 8-to-18 Year Olds*, Menlo Park, CA: Kaiser Family Foundation.

Rovio (2011) 'Angry Birds smashes half a billion downloads', available at: *http://www.rovio.com/en/news/blog/95/angry-birds-smashes-half-a-billion-downloads* (accessed 8 November 2012).

Ryan, R. M. (1982) 'Control and information in the intrapersonal sphere: An extension of cognitive evaluation theory', *Journal of Personality and Social Psychology* 43(3): 450–61.

Ryan, R. M. and Deci, E. L. (2000) 'Intrinsic and extrinsic motivations: Classic definitions and new directions', *Contemporary Educational Psychology* 25(1): 54–67.

Ryan, R. M. and Deci, E. L. (2002) 'Overview of self-determination theory: An organismic dialectical perspective', in E. L. Deci and R. M. Ryan (eds), *Handbook of Self-determination Research*, Rochester, NY: University of Rochester, pp. 3–33.

Ryan, R. M. and Grolnick, W. S. (1986) 'Origins and pawns in the classroom: Self-report and projective assessments of individual differences in children's perceptions', *Journal of Personality and Social Psychology* 50(3): 550–8.

Sharma, P. and Hannafin, M. (2007) 'Scaffolding in technology-enhanced learning environments', *Interactive Learning Environments* 15(1): 27–46.

Shuler, C. (2009) *Pockets of Potential: Using Mobile Technologies to Promote Children's Learning*, New York: The Joan Ganz Cooney Center.

Smith, A. (2010) 'Mobile access 2010', available at: *http://pewinternet.org/Reports/2010/Mobile-Access-2010/Summary-of-Findings.aspx* (accessed 8 November 2012).

Tran, C., Chen, J., Warschauer, M., Conley, A., and Dede, C. (2012) 'Applying motivation theories to the design of educational technology', in C. Martin, A. Ochsner and K. Squire (eds), *Proceedings of the Games, Learning, and Society Conference*, Pittsburgh, PA: ETC Press, pp. 291–7.

Traxler, J. (2007) 'Defining, discussing, and evaluating mobile learning: The moving finger writes and having writ...' *International Review of Research in Open and Distance Learning* 8(2): 1–12.

Trinder, J, Magill, J. and Roy, S. (2009) 'Using automatic logging to collect information on mobile device usage for learning', in G. Vavoula, N. Pachler and A. Kukulska-Hulme (eds), *Researching Mobile Learning*, Oxford: Peter Lang, pp. 241–56.

Vygotsky, L. S. (1978) *Mind in Society: The Development of Higher Psychological Processes*, Cambridge, MA: Harvard University Press.

Warschauer, M. (1996) 'Motivational aspects of using computers for writing and communication', in M. Warschauer (ed.),

Telecollaboration in Foreign Language Learning, Honolulu, HI: University of Hawai'i Second Language Teaching and Curriculum Center, pp. 29–46.

Warschauer, M. (1997) 'Computer-mediated collaborative learning: theory and practice', *The Modern Language Journal* 81(4): 470–81.

Warschauer, M. (2003) *Technology and Social Inclusion: Rethinking the Digital Divide*, Cambridge, MA: MIT Press.

Warschauer, M. and Grimes, D. (2007) 'Audience, authorship, and artifact: The emergent semiotics of Web 2. 0', *Annual Review of Applied Linguistics* 27(1): 1–23.

Webb, N. M. (1989) 'Peer interaction and learning in small groups', *International Journal of Educational Research* 13(1): 21–39.

Wilson, M. (2002) 'Six views of embodied cognition', *Psychonomic Bulletin & Review* 9(4): 625–36.

2

Using social network sites and mobile technology to scaffold equity of access to cultural resources

John Cook, Norbert Pachler and Ben Bachmair

Abstract: Using social media and personally-owned mobile devices as a means of providing a bridge from media use in everyday life to the expectations of school and higher education potentially has enormous attraction. This chapter discusses access to 'cultural resources' facilitated by digital media from a wide perspective (e.g. learning resources, health information, cultural events, employment opportunities, etc.). The purpose of the chapter is to reclaim the notion of 'cultural resources' for educational discourse as it seems powerful to us in the context of discussions about social mobility and fairness of access. Following a practical example which underscores the motivation of our discussion, key concepts are defined, particularly forms of 'capital' through the lens of the following question: how can we enable learning activities in formal and informal contexts undertaken by individuals and groups to become linked through scaffolding as a bridging activity mediated by network and mobile technology? As this is a large problem space, the literature is selectively reviewed in order to provide an overview of the current state of the art in terms of using network and mobile technology for bridging what we call 'social capital' (which emphasises the informational benefits of a heterogeneous network of weak ties). Tentative conclusions are drawn in an effort to further debate.

Keywords: social network site, mobile technology, cultural resources access, social mobility, social capital

Introduction

What motivates us to examine how network and mobile technology (NMT) can 'bridge' learning in formal and informal contexts? Few in the interdisciplinary field of research and development would disagree that it is a democratic right for citizens to have 'equity of access' to educational resources. NMTs offer unprecedented opportunities for communicating, interacting, sharing, meaning-making, content and context generation, etc. By focusing on the notion of 'cultural resources' in this chapter, we must first acknowledge that this term is often taken to be synonymous with 'cultural heritage' and 'high culture', i.e. museums, libraries, architecture etc. Indeed, a recent publication from the European Community Research and Development Information Service, views cultural resources in this way (CORDIS, 2011), while UNESCO (1972) provides a definition of 'cultural heritage' along similar lines. A different perspective (Kress, 2007) is one whereby cultural resources are viewed as shaping technology and its potential. And a slightly different perspective on cultural resources is to view technologies such as mobile devices and digital media, and the processes attendant to them and the affordances they offer, as cultural resources for learning:

> We propose a focus on the processes and artefacts, that is, on cultural resources, on the knowledge, skills and understandings that underpin an analytical engagement

> with media of all descriptions, but in particular with mobile devices, by individuals. This focus is premised on an examination of cultural resources and practices, an attempt to describe how cultural resources are distributed and an assessment of whether that distribution is 'fair'. Our aim is to support individuals in the use of such resources in their everyday life-worlds, which includes a critical stance on resources and practices where that seems called for. (Pachler, 2010)

Therefore, in this chapter we want to explore cultural resources defined broadly. We want to reclaim the use of the term 'cultural resources' as we see it closely linked to discussions about social mobility and equity of access. Specifically, our perspective acknowledges the fact that large segments of the population already have mobile devices that they own personally, and which they use in their everyday life-worlds, for meaning making and 'informal learning'. At the same time, not all citizens have equity of access to cultural resources (e.g. learning resources, health information, cultural events, employment opportunities, etc.). Thus it is the growing significance of mobile devices in learners' everyday lives, i.e. their ubiquity and personal ownership of them, as well as their increasing use for engaging with, and making sense of the world that motivates our interest in them.

The question begs asking why we use the adjectival qualifier 'cultural'. Why not just 'resources', particularly as 'cultural resources' has a lot of baggage, some of it political? Certainly we believe it is a democratic right to have equity of access to cultural resources (widely defined); indeed, there is some evidence that new digital media and devices enable at-risk learners and those not in education, employment or training (NEETs) to gain access to cultural resources (for example, see Learning and Skills Network, 2009; Spies and

Tredop, 2006). Nevertheless, we do not want to be prematurely convergent by, for example, looking simply at the knowledge economy. Instead, we want to encourage divergent thinking before – at a later date, and following debate – we can move to more convergent thinking.

Using social media and personally-owned mobile devices as a means of providing a bridge from widespread use in everyday life to the expectations of school and higher education has enormous potential. As we explore below, one approach to providing support is *scaffolding*,[1] which extends the socio-cultural approach of Lev Vygotsky.[2] Vygotsky's zone of proximal development (Vygotsky, 1978 [1930]) has formed the basis for much of the Learning Technology Research Institute's recent work, which the first author led until August 2012 (for examples of this work, see e.g. Cook, 2010; Smith et al., 2011), but the time has come to extend this in the context of enabling equitable access to cultural resources (broadly defined).

Social network sites (SNSs), a term mentioned in the chapter title, are:

> web-based services that allow individuals to (1) construct a public or semi-public profile within a bounded system, (2) articulate a list of other users with whom they share a connection, and (3) view and traverse their list of connections and those made by others within the system. The nature and nomenclature of these connections may vary from site to site. (Boyd and Ellison, 2008)

Furthermore, recent data (Mobithinking, 2010) has pointed to the trend whereby mobile users spend a lot more time on social networking sites than PC users. Yet, 'equity of access' to educational resources would, it is claimed here, also assume the ability of citizens to engage in the process of taking advantage of these technology-mediated opportunities. However, some issues have recently been noted (Merchant, 2011: 9) pertaining to the notion that the 'concept of a social network reduces the social human actor to a point – not even a point of view – but a point that connects in various ways to other points' and that 'there is much more depth and density of activity in social networks than is represented or enacted in SNSs'.

Mobile technology in the context of this chapter is best understood as mobile learning, which:

> as we understand it – is not about delivering content to mobile devices but, instead, about the processes of coming to know and being able to operate successfully in, and across, new and ever changing contexts and learning spaces. And, it is about understanding and knowing how to utilise our everyday life-worlds as learning spaces. Therefore, in case it needs to be stated explicitly, for us mobile learning is not primarily about technology. (Pachler et al., 2010: 6)

We also use the notion of 'life-worlds' in this chapter, which recognises the clustering of different factors such as socio-economic background, gender, age/generation, ethnicity, regional background, profession etc. Traditional boundaries of various kinds are being blurred and abolished, dissolving and disappearing. These trends have a lasting effect on meaning-making and learning.

This chapter takes into account the above perspectives and presents what is being claimed to be a new educational paradigm combining formal and informal learning processes centred on SNSs and mobile technology. Following the introduction, the chapter is structured as follows: a brief example of learning in everyday life-worlds is given to motivate the literature review and issues raised. The example is of resources for learning and draws on various traditions for its interpretation: philosophical traditions of idealism that take account of cultural resources (Humboldt, 2002); Leontjew (1973) a student of Vygotsky; and social class differences in the relevance of language to socialisation (Bernstein, 1987; Bernstein and Henderson, 1973). Cultural resources in the sense of idealism (Humboldt) or its materialist version (Leontjew) developed their education function by being appropriated. In our example below, the cultural resources are images from the internet, written texts of rap lyrics and the photos used in the teacher's mobile portfolio. Such artefacts become personal resources by being internalised and externalised (or represented) within the school context. We combine the dynamic of internalisation and externalisation with the term *appropriation*. Appropriation has three dynamic components: first, bringing cultural resources into a person's inner horizon of preferences, values, arguments or feelings etc.; second, processing, for example, the images from the internet; and, third, bringing out the results through expression within the context of the school. Furthermore, the following concepts are defined using the above key question as a lens: cultural capital (Bourdieu, 1986), social capital (Bourdieu, 1986; Putnam, 2001), informal learning in schools and higher education (Pachler and Cook, 2009), using mobile devices to bridge the gap between learning in formal and informal contexts (Cook et al., 2008) and scaffolding (Pol et al., 2010). Because this is a large problem space, the literature is then selectively reviewed to examine the current state of the art in terms of using NMT for scaffolding access to cultural resources (specifically: Livingstone and Brake, 2010; Steinfield et al., 2008; Tomai et al., 2010; Valenzuela et al., 2009). The chapter concludes by outlining the potentialities offered by NMT for sustainability in the integration of informal and formal institutional dimensions of learning. First, however, we give an example of cultural resources for learning in the form of an example on self-representation and writing in a school context.

Resources for learning: self-representation and writing in a school context

A recent case study from Germany (Bachmair 2012) focused on a cohort of 13-year-old boys with a migrant background at the lowest level of the tiered German school system. The boys were tasked with using the internet or their personal mobile devices to obtain material to produce a poster. They were also offered a short workshop on how to write and perform rap lyrics.

As discussed by Bernstein (Bernstein, 1987; Bernstein and Henderson, 1973), the verbal resources and symbolic forms of students from lower socio-economic groups are not recognised by schools as sufficiently elaborated for formal expressions in a curricular context. As such, the curricular design for the basic workshop on creative writing aimed to draw on the boys' own 'cultural resources' from their youth culture and everyday life.

From an educational and a curricular perspective, such a workshop entails two essential tasks. First, the tutor(s) of the workshop needs to recognise the cultural resources provided by the pupils as adequate and relevant material for self-expression. The second task is to integrate the pupils in a process of development so that their input has relevance for the school curriculum. Thus the goal is to lead the pupils from their naïve articulation of resources to a position where these resources are seen as necessary and relevant (for example, for getting an apprenticeship).

In the German study, the boys copied and pasted various materials from the internet such as strange male bodies, including a video of an extremely fat dancer, deformed babies and a baby with the head of a moustachioed old man. They did not store these images. The pupils engaged in a kind of 'consumptive appropriation' of cultural resources achieved by archiving, selecting, evaluating, transforming into a printable format and compiling images for their personal poster.

Figure 2.1 represents a visual summary of the process using images from the tutor's mobile portfolio (to preserve anonymity, pupils' faces have been rendered unrecognisable). One advantage of a mobile tutor portfolio is that the tutor can use the images as stimulus for reflection. As the cultural resources being employed by the tutor are similar to those being used by the pupils, it is interesting to compare the mobile portfolios of both parties (see Figure 2.2).

Figure 2.1 Images from the workshop tutor's mobile portfolio

Figure 2.2 Images from a pupil's mobile portfolio

With respect to the mobile portfolio resulting from the rap workshop, it was found that the pupils' photos emphasised the colour blue. Blue is a key design element of the local underground rail network, and thus thematically relevant for the boys, but not associated with a high level of reflection. Nonetheless, such a finding might be useful for identifying an intersection with the boys' creative expression. As expected, the boys used their mobile phones to represent themselves.

During the workshop, the tutor paid particular attention to the emergent writing process and the products of writing, the text of rap lyrics, and so on. However, this was not in the foreground of the pupils' interest. Instead, the pupils took photos of the sound studio's technical setup (Figure 2.3). The technical setup can thus be viewed as one of their resources of interest. Although this motivation was not unknown, it was now manifestly visible.

While the pupils' portrayals of the sound studio omitted people and pupils to focus on equipment such as the mixing desk, speakers, keyboards etc., the tutor's portfolio concentrated on the target-oriented performance of the pupils. This shows a striking difference in the evaluation or interpretation of the participants: while the tutor focused on

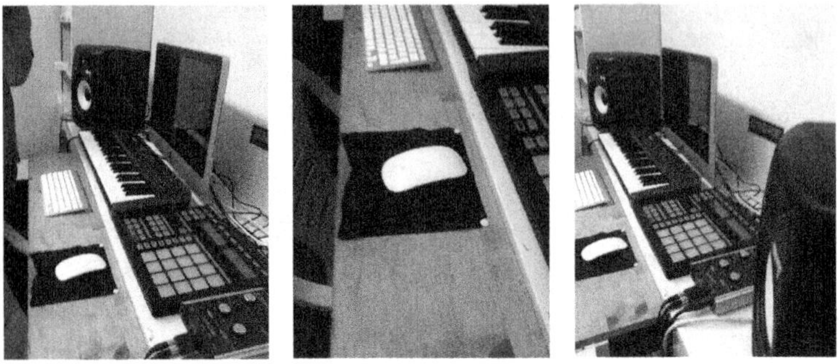

Figure 2.3 Sound studio portrayals

the studio–pupil interrelationship, the pupils foregrounded the studio's well-ordered technical setup as a cultural resource inviting them to be creative. The tutor's portfolio comprised the breadth of cultural resources within the workshop, ranging from copied internet images to a poster as a deliberate arrangement of images with direct relevance to the boys' age and gender-specific development. The tutor's image portfolio also showed the target-oriented written work of the pupils, recognising pupil achievement with respect to the traditional competence of creative writing.

Given the apparent disjuncture between the pupil and tutor view of motivating factors, we tentatively propose that the case study and associated analysis point to the need to re-examine the notion of cultural resources. Below we define the key concepts involved in such an undertaking.

Key concepts

In this section we define key concepts using the following question as a lens: how can we enable formal and informal learning activities of individuals and groups to become linked together through scaffolding as a bridging activity mediated by NMT?

Forms of capital

Although the London Mobile Learning Group (LMLG) did not derive the notion of cultural resources directly from Bourdieu, his work is nevertheless a relevant reference point, in particular his concept of 'cultural capital'. Bourdieu and Passeron (1979) argued that the cultural habits and dispositions of learners' families are fundamentally important for educational success and represent a resource, a form of cultural capital. From Bourdieu's perspective, schooling validates and augments cultural capital related to the learner's social background rather than develops it in those learners who come to school with limited resources.

Bourdieu (1986) distinguishes three states of cultural capital: the embodied, the objectified and the institutionalised. By embodied cultural capital he means dispositions of the mind and body such as those associated with the concept or habitus of *Bildung*;[3] by objectified cultural capital he means cultural goods, objects and media; and by institutionalised cultural capital he means objectifications conferred by institutional recognition such as academic qualifications.

The mobile complex of structures, agency and cultural practices of the LMLG's socio-cultural ecology (Pachler et al., 2010) can be seen to relate to Bourdieu's notion of cultural capital: cultural practices, such as the new habitus of learning developing around mobile devices and its services, can be seen to relate to embodied cultural capital, and the digital artefacts produced with and accessed through mobile devices can be understood as objectified cultural capital.

Furthermore, the LMLG also discuss the relationship of mobile cultural resources and education in our work with reference to empirical data on media use in everyday life in Germany (Bachmair, 2007) which inter alia show a social class bias tending to lead to the exclusion of children and young people from social groups with low income and at a distance to school-based education (this is reinforced by the above case study). The data also show that the ongoing process of individualisation and fragmentation is enforced by the media of everyday life, which function as cultural objects, as symbolic material integral to users' personal life-worlds. Additionally, the data show that socio-economic milieus frame users' media preferences and activity patterns.

Another view of 'capital' takes as a starting point the concept of social capital (e.g. Bourdieu, 1986; Putnam, 2001). For Bourdieu, social capital is:

> the aggregate of the actual or potential resources which are linked to possession of a durable network of more or less institutionalized relationships of mutual acquaintance and recognition – or in other words, to membership in a group – which provides each of its members with the backing of the collectively-owned capital. (Bourdieu 1986)

For Putnam (2001), and along seemingly similar lines to Bourdieu, social capital is the creation of social networks between socially heterogeneous groups; face-to-face examples are choirs and bowling clubs, online examples include the SNSs defined earlier. Putnam maintains that social capital is built in social networks and their associated norms of reciprocity. He distinguishes between bridging and bonding social capital. A useful overview of social capital has been provided by Tomai et al.

> Bridging social capital arises when people from various backgrounds make connections entering social networks that are 'inclusive' and therefore favour participation of individuals who differ on many crucial variables such as income, political orientation, ethnic origin, religious affiliation, etc. These kinds of networks literally create 'bridges', which is they allow people, who might not have had the possibility to encounter one another in their daily lives, the opportunity to become acquainted. The relationships that develop may lack in depth, but they offer breadth: they provide the chance to get to know people of manifold backgrounds. Alternatively, bonding social capital can be exclusive. It is built between individuals engaged in tightly-knit, emotionally close relationships, such as family and close friends. The individuals with bonding social capital have little diversity in their backgrounds, but have stronger personal connections. The continued reciprocity found in bonding social capital provides strong emotional and substantive support and enables mobilization. According to Putnam ... these two types of social capital are related but not equivalent. And moreover, they are not mutually exclusive. (Tomai et al., 2010: 265)

Thus, 'social capital is an elastic construct used to describe the benefits one receives from one's relationships with other people' (Steinfield et al., 2008: 434). In summary, bridging social capital is a term that emphasises the informational benefits of a heterogeneous network of weak ties whereas bonding social capital emphasises emotional benefits from strong ties to close friends and family (Steinfield et al., 2008: 435). This notion of weak ties provides a slight variation on the definition of social capital provided by Bourdieu in that if we allow weak ties, 'mutual acquaintance' in a network is not (necessarily) required. Furthermore, a key problem is one of creating and sustaining a spiral of social capital; such an endeavour may require scaffolding as a bridging activity (particularly for NEETs and at-risk learners). A key question is therefore: how can we enable formal and informal learning activities of individuals and groups to become linked together through scaffolding as a bridging activity mediated by NMT?

Informal learning in schools and higher education

We find the notion of 'informal learning' problematic at various levels, most specifically because we question whether fundamentally different cognitive (and social) processes are at work and, therefore, we prefer to use the term 'learning in informal contexts'. Notwithstanding these reservations, we recognise the fact that the term 'informal learning' is widely used and define it as follows:

> a natural activity by a self-motivated learner 'under the radar' of a tutor, individually or in a group, intentionally or tacitly, in response to an immediate or recent situation or perceived need, or serendipitously with the learner mostly being (meta-cognitively) unaware of what is being learnt. (Pachler and Cook, 2009: 152)

The literature reported below uses the term 'scaffolding' in various contexts, which is why we define this term here. The use of scaffolding as a metaphor refers to the provision of temporary support for the completion of a task that a learner might otherwise be unable to achieve. Pol et al.'s review of a decade of research on face-to-face scaffolding suggests that this work seems to 'point largely in the same direction, i.e. that scaffolding is effective' (Pol et al., 2010: 286). The conceptual state of the art in terms of face-to-face scaffolding is summarised diagrammatically in Figure 2.4. The figure represents a human tutor's external regulatory moves. Scaffolding social capital can potentially provide a link between an individual's self-regulatory moves and interactions with heterogeneous groups of people.

Cook et al. (2008) report on a study of the use of high-end mobile phones for off-site and on-campus mobile learning in the higher education (HE) setting.

The methodology used for the study draws on narrative and case-based approaches underpinned by grounded theory. As part of their formal assessment, university students were

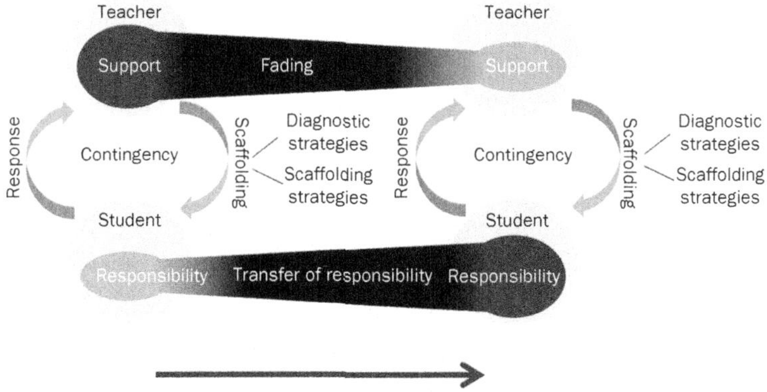

Figure 2.4 **Conceptual model of face-to-face scaffolding**

Source: Based on Pol, J., Volman, M. and Beishuizen, J. (2010) 'Scaffolding in teacher–student interaction: a decade of research', *Educational Psychology Review* 22(3): 271–96 (see p. 274).

given an assignment task which required them not only to gather data in the form of video clips and photos during field trips (i.e. learning in an informal context), but also to answer certain questions (i.e. fill knowledge gaps) that were posed by an events checklist (a 'bridge' provided by a mobile learning object installed on the phone). Qualitative data were collected through semi-structured interviews and analysed using a narrative case-study technique. Two dominant themes emerged from the cases: affective issues and phone usage in the learner's formal and informal practice.

Furthermore, the data suggest that in certain contexts, learner-sensitive scaffolding on the part of the tutor (e.g. via targeted text messages to students) can provide key support in terms of bridging the gap between learning in formal and informal settings. With respect to the early stages of learning, work on scaffolding by Lewin and Luckin (2010) has examined bridges between families and school. Two projects were examined by way of investigating the issues surrounding the use of technology to support parental involvement with schools and how children are learning. Findings suggest that technologies with readily accessible and flexible interactive resources can help develop parental engagement.

However, in relation to the 'bridging' that technology may provide between home and school, the study reports that:

> there are issues for both staff and parents. Without purposeful use, these challenges act as a barrier once they are past the novelty effect. We conclude that parental needs are complex and that engagement needs to be sensitively scaffolded rather than focusing on the technology. (Lewin and Luckin, 2010: 749)

Specifically, the system in this research was used to specify the way the resources were described, the design of the learner profiles and the way in which resources were combined into learning activities (Lewin and Luckin, 2010: 752).

The common theme emerging from the above studies in HE off-site and home-school learning in informal contexts is that scaffolding is needed to act as a bridge between informal and formal learning contexts but that it needs to be planned; consequently, we explore this notion in more detail below from the perspective of work on 'bridging social capital'.

Using NMT for bridging social capital

Because this is a large problem space, we review the literature selectively to examine the current state of the art in terms of using NMT for bridging social capital.

In many places, children are asked not to bring phones into school or the use of these devices within school is limited because of e-safety or harassment concerns (BBC, 2007). Schools are cautious – anxious, maybe – about new digital technologies and media. However, a comprehensive report from The Joan Ganz Cooney Center has examined over 25 mobile device oriented projects in the USA and throughout the world, reporting that there already exist 'early evidence and examples of how mobile devices may help redefine teaching and learning in the decade ahead' (Shuler, 2009: 4). Indeed, teachers can be seen as innovators, where the co-construction of learning between teachers and students is regarded as a powerful vehicle for innovation (Futurelab, 2007: 19). Our view is that it is very important for teachers in schools and HE to assume an obligation to see it as part of their role to sensitise young people to the need for a reflective use of new technologies particularly where bridging at-risk learners is concerned (i.e. providing guidance so that at-risk learners can move towards taking on the role of self-regulation in terms of digital literacy and 'safe' online behaviour). The issues surrounding bridging in HE are, naturally, rather different compared with in schools.

Main concerns in HE can be summarised in terms of challenges or disruption to the dominant virtual learning environment centric model that comes when organisations or individual learners and teachers take a more open approach where digital services, resources and devices can come from 'anywhere' (everyday life, convergent and individualised media). Indeed, there are concerns that student expectations about the provision of NMT in UK universities are not being met (Cooke, 2008). Consequently, given these emerging discourses it seems relevant and timely to look at the use of NMT, in particular, for attempts to scaffold individuals' self-regulatory abilities and link them to the social capital provided by heterogeneous networks (we call this bridging social capital for short).

Given our stated interest in at-risk learners, the work of Livingstone and Brake (2010) is relevant here. These authors have reviewed recent findings regarding children and teenagers' social networking practices in order to:

> identify implications for future research and public policy. These focus on the interdependencies between opportunities and risks, the need for digital or media literacy education, the importance of building safety considerations into the design and management of social networking sites, the imperative for greater attention to 'at risk' children in particular, and the importance of a children's rights framework in developing evidence-based policy in this area. (Livingstone and Brake, 2010: 75)

Children who may be 'at risk' of victimisation, typically have low self-esteem or lack satisfying friendships or may even perpetrate harm towards others. Livingstone and Brake propose that:

> balanced risk assessment should also note that, though dangerous, risks to children from adult sexual predators on social networking sites are very rare ... and more common is the misuse of personal information by spammers and fraudsters ... and the inadvertent release of personal information harmful to young people's reputations and employment prospects. (Livingstone and Brake, 2010: 80)

These researchers also point out that:

> [the] balance between opportunities and risks should, arguably, be struck differently for 'at risk' children, where greater monitoring or restrictions may be legitimate – moreover, for these children especially, relying on parents to undertake this role may be inappropriate. (Livingstone and Brake, 2010: 80)

A study by Tomai et al. (2010) reports that 'levels of bridging and bonding social capitals were significantly higher for members [of online networks, but that] ... intensity of use was significantly related only to bridging but not to bonding social capital'. Tomai et al. report that 'trust' is a key issue for social capital:

> Members of the online community had more sense of belonging to their school, they perceived themselves more as members of a wider community, stated they were more interested in their school and more willing

> to support school's activities. Our data show that members had also significantly higher levels of bonding social capital ... Members reported more perceived emotional support from their relations than non-members: a sign that their social bonds were fairly intimate and not superficial. Members stated more often that in their school they had people they trusted who could help them solve their problems, or people whom they could asked for advice on important matters. Non-members declared more often that nobody was available at school to talk about delicate matters, while members stated more often that when they felt lonely they had people with whom to talk. Moreover members seem to perceive they could get also more types of material support from their schoolmates than non-members, such as getting some money for an emergency. (Tomai et al., 2010: 272)

Mixed results on bridging social capital in HE were found in a longitudinal study in the USA, which investigated the relationship between intensity of Facebook use, measures of psychological well-being, and bridging social capital:

> Two surveys conducted a year apart at a large US university, complemented with in-depth interviews with 18 Facebook users, provide the study data. Intensity of Facebook use in year one strongly predicted bridging social capital outcomes in year two, even after controlling for measures of self-esteem and satisfaction with life. These latter psychological variables were also strongly associated with social capital outcomes. Self-esteem served to moderate the relationship between Facebook usage intensity and bridging social capital: those with lower self-esteem gained more from their

> use of Facebook in terms of bridging social capital than higher self-esteem participants. (Steinfield et al., 2008)

The researchers go on to suggest that Facebook affordances 'help reduce barriers that lower self-esteem students might experience in forming the kinds of large, heterogeneous networks that are sources of bridging social capital' (Steinfield et al., 2008). In contrast, another study of Facebook among college students in the USA suggested that:

> [although the] findings should ease the concerns of those who fear that Facebook has mostly negative effects on young adults, the positive and significant associations between Facebook variables and social capital were small, suggesting that online social networks are not the most effective solution for youth disengagement from civic duty and democracy. (Valenzuela et al., 2009)

Conclusions

The introduction claimed that the chapter would present a new educational paradigm combining formal and informal learning processes centred on SNSs and mobile technology. However, our case study in schools and our review of the reported research in HE showed mixed results. Some research suggests that in HE Facebook, for example, provides affordances that can help reduce barriers that students with lower self-esteem might experience in forming the kinds of large, heterogeneous networks that are sources of social capital. 'Trust' is a key issue in this respect. Thus, there appears to be considerable potential for NMT in terms of sustainability in the integration of informal and formal institutional dimensions of learning. However, although a new educational paradigm is emerging, there exists a need for more debate and further research. As it stands, there still appears to exist a small conceptual gap around cultural resources.

Society and cultural forces help shape technology, and in this sense it has been said that 'we cannot jump over our shadows' (Kress, 2007). Do such forces also set the limits of appropriation and transformation of technology and indeed define the level of access to cultural resources? We cannot jump over our shadows (and hence 'appropriate' the shadow, i.e. make use of shadows in a culturally novel way to meet a new purpose) – or can we? Extending the shadow metaphor, the principles of sundials can be understood better from the perspective of the sun's apparent motion across the sky. In fact, a sundial is a latitude-specific technology to indicate the time; it uses (or appropriates) the shadow created by the sun's light as the Earth spins on its axis around the sun. The object casting the shadow is the sundial's gnomon, commonly a triangular blade on top of the sundial. Maybe there are occasions when we can jump over our shadows by appropriation, or when we need help to see beyond. In this sense, innovative and creative thinkers and groups have always been able to jump over the shadow created by society and culture. Is that not a paradigm shift? Or is it a creative act – a reaction propelling a concept over the historical shadow of society? On a more personal level, what if we want to provide equity of access to cultural resources for individuals and groups? What do we mean by this and how could we achieve it? This chapter has started to explore these issues but our work over the coming years will attempt to address these challenging issues in even more depth.

Notes

1. The use of *scaffolding* as a metaphor refers to the provision of temporary support for the completion of a task that a learner might otherwise be unable to achieve.
2. Of course, Vygotsky never used this term – it was coined by Wood et al. (1976).
3. The term *Bildung* refers to the German tradition of self-cultivation, wherein philosophy and education are linked in manner that refers to a process of both personal and cultural maturation. This maturation is described as a harmonisation of the individual's mind and heart and in a unification of selfhood and identity within the broader society.

Bibliography

Bachmair, B. (2007) 'M-learning and media use in everyday life: towards a theoretical framework', in N. Pachler (ed.), *Mobile Learning: Towards a Research Agenda*, Occasional Papers in Work-based Learning Vol. 1, London: WLE Centre, pp. 105–52.

Bachmair, B. (2012) 'Kulturressourcen für Selbstrepräsentation jugendlicher Migranten im Kontext der Schule', in A. Kriwak and G. Pallaver (eds), *Medien und Minderheiten*, Innsbruck: Innsbruck University Press, pp. 173–98.

BBC (2007) 'School mobile phone ban ruled out', available at: *http://news.bbc.co.uk/1/hi/scotland/glasgow_and_west/6334523.stm* (accessed 3 September 2012).

Bernstein, B. and Henderson, D. (1973) 'Social class differences in the relevance of language to socialization', in B. Bernstein (ed.), *Class, Codes and Control. Volume 2: Applied Studies towards a Sociology of Languages*, London and Boston, MA: Routledge & Kegan Paul, pp. 24–47.

Bernstein, B. (1987) *Elaborated and Restricted Codes: An Overview, 1958–1986*, Occasional Papers Vol. 2, Amsterdam: Amsterdam University, Centre for Race & Ethnic Studies.

Bourdieu, P. (1986) 'The forms of capital', in J. G. Richardson (ed.), *Handbook of Theory and Research for the Sociology of Education*, Westport, CT: Greenwood Press, pp. 241–58.

Bourdieu, P. and Passeron, J.C. (1979) *The Inheritors: French Students and their Relations to Culture*, Chicago, IL: University of Chicago Press.

Boyd, D. M. and Ellison, N. B. (2008) 'Social network sites: definition, history, and scholarship', *Journal of Computer-Mediated Communication* 13(1): 210–30.

Community Research and Development Information Service (CORDIS) (2011) 'ICT Challenge 8: ICT for learning and access to cultural resources', available from: *http://cordis.europa.eu/fp7/ict/programme/challenge8_en.html* (accessed 3 September 2012).

Cook, J. (2010) 'Mobile phones as mediating tools within augmented contexts for development', *International Journal of Mobile and Blended Learning* 2(3): 1–10, available at: *http://www.igi-global.com/journal/international-journal-mobile-blended-learning/1115* (accessed 11 November 2009).

Cook, J., Pachler, N. and Bradley, C. (2008) 'Bridging the gap? Mobile phones at the interface between informal and formal learning', *Journal of the Research Center for Educational Technology* 4(1): 11–17.

Cooke, R. (2008) *On-line Innovation in Higher Education*, Submission to the Rt Hon John Denham MP, Secretary of State for Innovation, Universities and Skills, 8 October.

Futurelab (2007) 'Teacher innovation: A look at the possibilities for innovation within the current education framework', *Vision*, No. 4, available at: *http://www.futurelab.org.uk/resources/documents/vision/VISION_04.pdf* (accessed 4 March 2010).

Humboldt, W. von (2002 [1792]) 'Ideen zu einem Versuch, die Gränzen der Wirksamkeit des Staates zu bestimmen', in A. Flitner and K. Giel (eds), *Wilhelm von Humboldt. Werke in fünf Bänden. Band I: Schriften zur Anthropologie und Geschichte* (4th edn), Stuttgart: Wissenschaftliche Buchgesellschaft, pp. 56–233.

Kress, G. (2007) 'Culture, technology and (environments of) learning', *Ideas in Cyberspace Education*, available at:

http://www.education.ed.ac.uk/ice3/papers/keynote_kress.html (accessed 1 March 2007).

Learning and Skills Network (2009) *Tackling the NEETs Problem. Supporting Local Authorities in Reducing Young People Not in Employment, Education and Training*, London: Learning and Skills Network.

Leontjew, A.N. (1973) *Probleme der Entwicklung des Psychischen*, Frankfurt: Volk und Wissen.

Lewin, C. and Luckin, R. (2010) 'Technology to support parental engagement in elementary education: Lessons learned from the UK', *Computers & Education* 54(3): 749–58.

Livingstone, S. and Brake, D. R. (2010) 'On the rapid rise of social networking sites: new findings and policy implications', *Children & Society* 24(1): 75–83.

Merchant, G. (2011) 'Unravelling the social network: theory and research', *Learning, Media and Technology* 37(1): 4–19.

Mobithinking (2010) 'Mobile social networking – the statistics are compelling', available at: *http://mobithinking.com/blog/mobile-net-stats* (accessed 8 June 2009).

Pachler, N., Bachmair, B. and Cook, J. (2010) *Mobile Learning: Structures, Agency, Practices*, New York: Springer.

Pachler, N. and Cook, J. (2009) 'Mobile, informal and lifelong learning: a UK policy perspective', in K. Nyíri (ed.), *Mobile Communication and the Ethics of Social Networking*, Vienna: Passagen Verlag, pp. 149–57.

Pol, J., Volman, M. and Beishuizen, J. (2010) 'Scaffolding in teacher–student interaction: a decade of research', *Educational Psychology Review* 22(3): 271–96.

Putnam, R.D. (2001) *Bowling Alone: The Collapse and Revival of American Community*, Vol. 108, New York: Simon & Schuster.

Shuler, C. (2009) *Pockets of Potential: Using Mobile Technologies to Promote Children's Learning*, New York: The Joan Ganz Cooney Center at Sesame Workshop.

Smith, C., Bradley, C., Cook, J. and Pratt-Adams, S. (2011) 'Designing for active learning: putting learning into context with mobile devices', in A. D. Olofsson and J. O. Lindberg (eds), *Informed Design of Educational Technologies in Higher Education: Enhanced Learning and Teaching*, Hershey, PA: IGI Global pp. 307–29.

Spies, A. and Tredop D. (2006) *Risikobiographien. Benachteiligte Jugendliche zwischen Ausgrenzung und Förderprojekten*, Wiesbaden: Verlag.

Steinfield, C., Ellison, N. and Lampe, C. (2008) 'Social capital, self-esteem, and use of online social network sites: A longitudinal analysis', *Journal of Applied Developmental Psychology* 29(6): 434–45.

Tomai, M., Rosa, V., Mebane, M. E., D'Acunti, A., Benedetti, M. and Francescato, D. (2010) 'Virtual communities in schools as tools to promote social capital with high schools students', *Computers & Education* 54(1): 265–74.

United Nations Educational, Scientific and Cultural Organisation (1972) 'Convention concerning the protection of the world cultural and natural heritage', available at: *http://whc.unesco.org/archive/convention-en.pdf* (accessed 3 September 2012).

Valenzuela, S., Park, N. and Kee, K. F. (2009) 'Is there social capital in a social network site? Facebook use and college students' life satisfaction, trust, and participation', *Journal of Computer-Mediated Communication* 14(4): 875–901.

Vygotsky, L. (1978 [1930]) *Mind in Society. The Development of Higher Psychological Processes*, edited by M. Cole, V. John-Steiner, S. Scribner and E. Souberman, Cambridge, MA: Harvard University Press.

Wood, D., Bruner, J. and Ross, G. (1976) 'The role of tutoring in problem solving', *Journal of Child Psychology and Psychiatry and Allied Disciplines* 17: 89–100.

3

A mobile Web 2.0 framework: reconceptualising teaching and learning

Thomas Cochrane and Roger Bateman

Abstract: This chapter presents a mobile Web 2.0 framework for pedagogical change based upon the implementation of over 30 mobile learning (m-learning) projects between 2006 and 2011. These projects explored the potential of m-learning as a catalyst for pedagogical change within architecture, product design, landscape architecture, contemporary music, computing, graphics design, performing and screen arts, accountancy law and finance, civil engineering, and journalism. The projects utilised mobile devices for student-generated content and for enabling student-generated learning contexts, bridging situated experiences with the formal classroom. Examples of the development and implementation of the framework are drawn from the context of the bachelor of product design programme at Unitec, New Zealand. The framework is founded upon contemporary social learning theory and illustrates the potential of mobile Web 2.0 tools to bridge pedagogically designed learning contexts between formal and informal learning and across international boundaries.

Keywords: m-learning, pedagogy-andragogy-heutagogy continuum, mobile Web 2.0 tools, pedagogical change, pedagogy 2.0

Introduction

This section outlines the authors' experiences of implementing mobile Web 2.0 projects from 2006 to 2011, the main drivers, and the key impacts on teaching and learning achieved, in particular the ability to bridge the formal and informal learning contexts.

The research was driven by a desire to bring about positive pedagogical change, informed by this research, in the areas of professional development for lecturers to utilise and integrate mobile Web 2.0 tools into their curricula to support flexible social constructivist learning environments, while facilitating the changes in institutional strategy and wireless infrastructure required to foster a student-owned wireless mobile device model of computing.

Several factors contributed to make this a possibility: the roll-out of almost ubiquitous wireless connectivity via Wi-Fi and 3G broadband, the maturing of smartphones into powerful mobile multimedia computers with unique affordances to augment how we conceptualise and interact with the world around us, the rapid development of mobile Web 2.0, and the development of new social constructivist pedagogies such as authentic learning (Herrington and Herrington, 2006; Herrington and Oliver, 2000), pedagogy 2.0 (McLoughlin and Lee, 2008, 2010), connectivism (Siemens, 2004) and navigationism (Brown, 2005, 2006).

The authors formed a collaborative partnership beginning in 2006 with an initial m-learning pilot project within the bachelor of product design (BPD) programme at Unitec. Based in Auckland, Unitec is New Zealand's largest institute of technology, with 600 lecturers and 10,000 students studying courses ranging from vocational certificates to doctoral post-graduate degrees. Based upon a participatory action research methodology (Swantz, 2008) the authors' m-learning projects led to the development of an implementation framework across the three years of the product design degree programme in 2009 and 2010 (Cochrane and Bateman, 2011a).

This research methodology was very similar to the design-based research methodology that Reeves (2005) advocates for transformational educational technology research. These experiences were further built upon in 2011 in the context of an international collaborative project exploring the potential pedagogical impact of mobile Web 2.0 tools across the contexts of six different learning communities in four countries across the globe (Cochrane et al., 2011).

The present authors' collaborative m-learning projects have included:

- *2006*: An initial m-learning trial among third-year BPD students using Palm TX and LifeDrive personal digital assistants (PDAs).
- *2007*: Instigation of third-year BPD students' Web 2.0 e-portfolios.
- *2008*: Voluntary m-learning projects with first, second and third-year BPD students, using Nokia N80, N95 and iPhone 3G handsets and Web 2.0.
- *2009*: Integration of mobile Web 2.0 e-portfolios across the entire BPD course, using Nokia XM5800, N97 and Dell Mini9 3G netbooks.
- *2010*: Building on 2009, but with a focus upon student-owned devices and collaborative projects with other student groups around the world.
- *2011*: Establishment of an international collaborative m-learning project involving four countries and six student groups.

Beyond the context of product design, the researchers have also facilitated wireless mobile device (WMD) m-learning projects within a variety of other tertiary education contexts, as summarised in Figure 3.1.

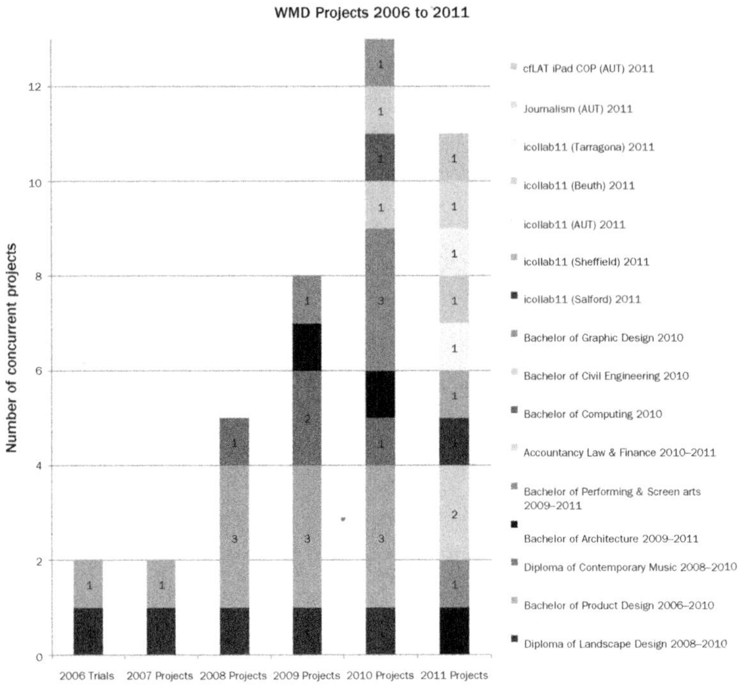

Figure 3.1 The authors' m-learning projects 2006 to 2011

Background

Several theoretical foundations were used to inform the implementation framework and practice, and are outlined in the following section.

Social learning theory

The researchers chose social constructivism (Vygotsky, 1978) as the foundational learning theory on which to base the m-learning projects. As such, pedagogical frameworks that emphasise social learning and collaboration informed the projects' design and assessment decisions. These included the following pedagogical frameworks: pedagogy 2.0, student-generated contexts, communities of practice (COPs), and authentic learning.

Pedagogy 2.0

Pedagogy 2.0 (McLoughlin and Lee, 2007, 2008, 2010) involves the appropriation of Web 2.0 tools to enable new and emerging pedagogies such as communities of practice, authentic learning and learner-generated-contexts. Key aspects of pedagogy 2.0 involve the integration and modelling of the use of mobile and Web 2.0 technologies in educational contexts.

Learner-generated contexts

Learner-generated contexts and the pedagogy-andragogy-heutagogy (PAH) continuum (Cook, 2010; Garnett, 2010; Luckin et al., 2010) focus upon transforming teaching from teacher-directed pedagogies to learner-directed or negotiated activities, empowering learners to become self-directed learners. Learner-generated contexts bridge the typical divide between teacher-directed pedagogies (pedagogy) associated with lower-level education, student-centred pedagogies (andragogy) and the student-directed learning pedagogies (heutagogy) usually the domain of higher education such as doctoral studies.

We describe the process of reconceptualisation of teaching and learning from pedagogy to heutagogy as an ontological shift (Chi and Hausmann, 2003) for the participants: as based upon their previous educational experiences, the lecturers reconceptualise what it means to teach, and students reconceptualise what it means to learn within social constructivist paradigms. The m-learning projects have illustrated that these ontological shifts require significant time and scaffolding, which can be achieved by framing courses around COPs.

Communities of practice

'Communities of practice' is a social learning theory (Lave and Wenger, 1991; Wenger et al., 2005, 2009). COPs can provide a framework for developing and nurturing learning communities. Central to the development of COPs are the concepts of legitimate peripheral participation, technology stewardship, and brokering of the boundary objects that represent the reified activities of the COP beyond the confines of the COP.

The goal of a COP is to bring the peripheral members of the community into full participation within the community guided by expert (or more experienced) peers who form the core group of the COP. This is similar to the concept of the zone of proximal development (Vygotsky, 1978; Bryant, 2006). The activities of the m-learning project COPs are physically represented in participant-generated artefacts such as YouTube videos, blog posts and Twitter streams. These artefacts were then used by the authors to model and broker the impact of this approach to other lecturers and students across the institution and internationally.

Authentic learning

Authentic learning (Herrington and Herrington, 2007; Herrington et al., 2009) provides a methodology for creating and supporting learner-generated content and contexts. Authentic learning is based on constructivist learning paradigms, and specifically on situated learning theory, focusing upon bridging theory and practice in learning. In the authors' view, the ability of m-learning to bridge learning contexts and facilitate student-generated contexts provides a powerful way to address the critics of authentic learning who contend that situated learning cannot authentically occur in the classroom. Networked mobile technologies (NMTs) can facilitate the design of pedagogical activities that enable students to continue learning conversations and experiences both in and beyond the classroom.

M-learning

Mlearning is a catalyst for pedagogical change that can be leveraged by lecturers modelling the pedagogical use of mobile Web 2.0 tools for facilitating reflective reconception of teaching and learning, moving from teacher-directed pedagogy to learner-generated content and learner-generated contexts. The introduction of m-learning into a course can act as a catalyst for pedagogical change by disrupting existing power relationships within the teaching and learning environment, creating a foundation for facilitating student-generated content and student-generated contexts. Thus, a move from pedagogy (lecturer-directed and lecturer-generated content) to andragogy (student-centred and student-generated content) and finally heutagogy (student-directed or student-negotiated and student-generated contexts) can be facilitated. The 2010 JISC mobile review concludes that mobile learning presents the potential to drive innovation in education:

> Mobile learning may mean different things to different people, but it is the dialogue that an institution begins with itself, its staff, its learners, its community – that

> matters. It is certainly not time for 'business as usual'. It is time to define and start driving innovation. (Belshaw, 2010: 63)

M-learning has moved beyond the realms of fantasy to become a viable platform for contextual learning that bridges formal and informal learning environments in and beyond the classroom. Kukulska-Hulme emphasises the potential impact of m-learning:

> With its strong emphasis on learning rather than teaching, mobile learning challenges educators to try to understand learners' needs, circumstances and abilities even better than before. This extends to understanding how learning takes place beyond the classroom, in the course of daily routines, commuting and travel, and in the intersection of education, life, work and leisure. (Kukulska-Hulme, 2010: 181)

Pedagogical transformations through m-learning

The ability of m-learning to act as a catalyst for pedagogical change is of great interest to the present authors. Indeed, it formed the basis for developing a design framework for m-learning that is founded upon a social constructivist pedagogy that enables learner-generated content and learner-generated contexts that bridge both formal and informal learning environments. Thus, rather than being technology-centric, the impact of m-learning can be seen in terms of its potential for pedagogical transformation. We are therefore in agreement with Pachler et al.:

> Mobile learning – as we understand it – is not about delivering content to mobile devices but, instead,

> about the processes of coming to know and being able to operate successfully in, and across, new and ever changing contexts and learning spaces. And it is about our understanding and knowing how to utilise our everyday life-worlds as learning spaces. Therefore, in case it needs to be stated explicitly, for us mobile learning is not primarily about technology. (Pachler et al., 2010: 6)

Thus, our focus has been upon exploring the potential for pedagogical transformation enabled by the educational affordances of smartphones as the device that students already own. Smartphones are a subset of the NMTs referred to in this book, emphasising the wireless connectivity and communication affordances of these technologies. The potential of these devices to enable the design of new pedagogical strategies that bridge formal and informal learning are illustrated in the BPD examples that follow.

Mobile Web 2.0 design framework: some examples of use

A mobile Web 2.0 framework was developed across several iterations of m-learning project implementation within the context of the BPD course at Unitec, New Zealand. This framework was subsequently transferred to an international collaborative project between six different courses in four different countries and implemented in 2011. Thus, these examples demonstrate the use of a mobile Web 2.0 design framework to bridge multiple learning contexts (including formal and informal) and reconceptualise teaching and learning.

The BPD programme, 2006 to 2011

In 2006, when the BPD programme initially explored the mobile Web 2.0 framework, it was a three-year undergraduate degree programme of 360 credits of full-time study. The programme was launched in 2003 and was borne out of a previous bachelor's programme that had its roots in a traditional approach to design studio teaching that favoured the atelier or 'private method' of instruction, where an individual staff member works with a small group of students to train them progressively.

Art and design education has frequently argued that the atelier method or studio environment, mirrors the 'real world'; however, the intellectual landscape of the twenty-first century, made up of emergent organisations, open source development and networked innovation, rises from new and constantly changing human connections.

The landscape of work environments has also been changing to strengthen these connections and give people more variety and choice in where to work and how to work. Technology and the internet allow constant access and wider access than ever before.

The standard atelier method or studio teaching environment of one communal space and one fixed timetable is unlikely to offer the best support and learning opportunities for today's creative students; it does not mirror the 'real contemporary world'.

Over a four-year period, the integration of mobile Web 2.0 tools into the BPD course facilitated significant flexibility for students, allowing them to stay connected, share their ideas widely, participate in worldwide creative communities and choose to work in virtually any context on or off campus.

2006

Beginning in 2006, an m-learning trial using Palm Wi-Fi PDAs and Web 2.0 tools such as Blogger.com and instant messaging was implemented within one project in the third year of the BPD programme (Cochrane, 2006). This was the first attempt at integrating m-learning within the course. However, there was little course integration, limited buy-in from course lecturers, limited campus Wi-Fi coverage, and the results effectively illustrated how not to approach m-learning. At the same time, the lead researcher was developing a COP model for educational technology literacy in tertiary academics (Cochrane and Kligyte, 2007).

Product design course lecturers were invited to form an intentional COP (Langelier, 2005) to investigate the use of Web 2.0 tools within their teaching. This first attempt at establishing a lecturer COP was short-lived, although one lecturer was motivated to explore these ideas further in 2007. While no formal changes were made to the traditional paper-based implementation of the major project in 2006, reflections on these experiences merged to form the foundational concepts underpinning subsequent implementation and research into mobile learning. The 2006 trials were also used to develop and test the research questions and data collection instruments.

2007

In 2007, the main third-year lecturer decided to integrate the optional use of Web 2.0 tools such as blogging (via WordPress) into the third-year course using student-owned laptops and desktops. This integration was achieved with regular technological support from the lead researcher. Significant advantages in moving to this learning environment were envisioned by the lecturer:

> Research shows us that there are 'far more dyslexic Art and Design students than we ever realised' (Hercules, 2001: 2) and that dyslexia raises many issues for studio-based teaching methodologies. By implementing the use of student reflective design journals as living, media-rich blogs it was hoped that these students would be engaged and empowered in their learning. (Course lecturer, 2007)

This was achieved by modifying the core assessment of the third-year programme that focused upon three student-defined product designs throughout the entire year. The impact of this pedagogical intervention on the teaching and learning environment is summarised by the lecturer below:

> Thinking about what for us as designers and training young designers – what is 'real world learning'? Real world learning involves team working, and blogs allow you to work in teams in a way that you can't work if you don't use them. We see the use of blogs as a way of being able to stay in touch in a kind of multi-lane highway – rather than a single stream. It's something that's allowed staff to engage with students in a way that doesn't happen with e-mail and so on. In terms of our profession it's absolutely vital that we do this – and I'm keen to sit down with my colleagues and see how we can embed this into the programme rather than in a particular year of the programme – and we can get the students from first, second and third year interfacing with each other and their blogs. (Lecturer, 2007)

To support the 2007 work, a product design lecturer COP was set up to investigate the integration of Web 2.0 and mobile Web 2.0 into the course in the second half of 2007. The lecturer

COP was then used as a model for supporting students in a COP that comprised the researcher as the technology steward (Wenger et al., 2005), the course lecturers and the students.

2008

In late 2007, the goals for 2008 were set out. The focus was on an m-learning integration plan that would utilise a range of mobile Web 2.0 tools and software with the BPD students along with a range of assessment criteria including podcasting and vodcasting. The hope was that those students who underperformed due to literacy problems would find a 'natural' way to blog their projects and thus develop the desired critical reflective thinking skills.

A project outline for 2008 was developed by the researchers and presented for discussion with the other course lecturers. The 2008 m-learning project budget allowed for students to be supplied with smartphones and Bluetooth folding keyboards. Students were encouraged to use them as if they owned the devices. Students were also supplied with a 1 GB data plan for the duration of the course. It was our opinion that programmes need to provide the hardware or make purchase a compulsory course-related cost.

The initial stages of the 2008 project saw a drop-off in 'normal' project activity as students explored the mobile Web 2.0 tools, including setting up the software and hardware and having fun exploring the new technology that was available to them. However, as the tools became integrated into the students' daily workflows, a significant uptake in engagement in the course was observed. This formed the foundation of student-generated e-portfolios that students could use to capture and share their learning experiences in virtually any context, both formal and informal, using their mobile devices.

2009

In 2009, a further m-learning project iteration was rolled out for the duration of the third year of the course. In 2009, students were required to integrate Web 2.0 into their studio practice. To this end, the programme provided Nokia N95 smartphones and a weekly COP meeting that focused on understanding and experimenting with Web 2.0 tools and technologies. Whereas in 2008 students were given a 1 GB data plan for the duration of the course, in 2009 they funded the data and voice connection plans themselves but were provided free Wi-Fi access while on campus.

SHaC09

The Sustainable Habitat Challenge (SHaC) is a national competition in the form of a collaborative project for teams around New Zealand to design, develop, and build sustainable housing in their local community. Throughout the SHaC09 project, data-sharing was enabled through a range of software applications. Staff and students made project work and resources available to the rest of the world online, via blogs, wikis and other Web 2.0 applications. Figure 3.2 illustrates the range of Web 2.0 tools used and the types of interaction between the project members facilitated by these tools.

Good project management, collaborative working and cross-departmental communications were vital to the success of the project. Within the context of SHaC09, departmental-specific academic briefs were developed collaboratively by lecturers in the departments of design, landscape design, communication and applied trades.

Web 2.0 tools including Vox, Ning and Flickr were used to develop the briefs and supplement in-person meetings during the writing stage. Product design students worked in

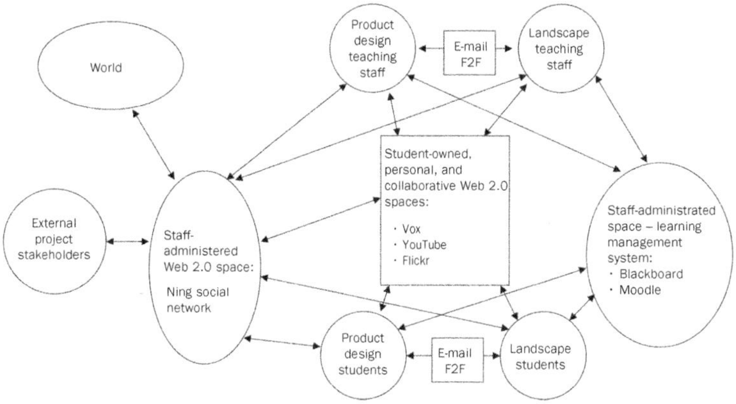

Figure 3.2 SHaC09 mobile Web 2.0 network

one of five groups, each of which focused on a specific design challenge. Students were required to carry out aspects of research in their group, sharing information via group meetings and Web 2.0 tools (Cochrane et al., 2009a). This project thus explored the potential of mobile Web 2.0 tools to enable and enrich student collaboration across departmental boundaries.

The nomadic studio

Social software tools can be effectively integrated into both face-to-face and online environments. The most promising settings for a pedagogy that capitalises on the capabilities of these tools are fully online or blended so that students can engage with peers, instructors, and the community in creating and sharing ideas (McLoughlin et al., 2008: 3). Moving further away from the atelier method as outlined above and building upon the work carried out in 2008, our research focus for 2009 focused on the seamless integration of Web 2.0 into the BPD course as well as augmenting the level of flexibility for students to allow them to work in virtually any context on or off campus.

In the second semester of 2009, students were required to undertake a regular 'nomadic' session where they would work away from the studio, but continue collaborating and learning conversations via mobile Web 2.0 connectivity (Cochrane et al., 2009a).

During the 'nomadic' studio session, students were expected to:

- be online via MSN or following their lecturer and classmates on Twitter;
- make at least one relevant blog post summarising their work;
- upload some multimedia content capturing what they were doing – for example a Qik or Flixwagon videostream or recorded vodcast, or to geotag and upload a photo to Flickr, Picasa or similar.

Thus, the nomadic studio formed a virtual bridge between the formal physical studio context and the informal situated learning contexts in which the students chose to work and research.

2010

In 2010, the researchers' aim was to embed the Web 2.0 framework into the BPD course as a whole. The researchers planned a strategy to work with the first-year students as they had not previously introduced the first-years to the mobile Web 2.0 framework. The first-year implementation focused on the first stage in this transformation, facilitating student-generated content and collaboration (Cochrane and Bateman, 2010b). Examples of assessment alignment and integration of the mobile Web 2.0 tools within the course are outlined below.

Students and lecturers were provided with Dell Mini9 Wi-Fi and 3G capable netbooks for the first semester. At the end of the first semester, the students and lecturers were also provided with a Wi-Fi and 3G capable smartphone that integrated a 3.2 megapixel camera, video recording, GPS, touch-screen for text input, and multitasking operating system for a variety of Symbian based applications. Students and lecturers were encouraged to personalise the use of these mobile devices and treat them as if they owned them for the duration of the year.

The first-year m-learning projects focused primarily on students establishing and personalising the use of core Web 2.0 tools that could then be built upon more explicitly in the second year of the course where the focus moved to mobile-specific affordances. The focus was therefore more upon the use of the netbook than the smartphones during the first-year course, establishing Web 2.0 e-portfolios for the students that would be built upon in 2010. The following sections outline the first year projects.

Semester 1: Ergonomics assignment

The goal of this assignment was for students to take into account the product user, the physical product and the context of use within a product design project. The project was designed to give students an introduction to conducting a controlled research project and prototyping test rigs to measure quantitative and qualitative data. Students used their netbook to create and establish an online journal/blog of their design investigation via the now defunct Vox blogging service. Students invited their peers, lecturers and the researchers (including the technology steward) into their 'neighbourhood' to facilitate sharing, commenting and critiquing, creating a virtual collaborative learning environment to augment the traditional face-to-face studio environment. Thus, students used Web 2.0 tools for social collaboration within their course, but also had the opportunity to share this process and content with a potentially worldwide audience. These blogs then become the core of the students' online e-portfolio that will be developed over the next three years of their course.

Semester 2: Practice in context 2, assignment 1

Practice in context is a theory-based course that introduces students to key exponents in contemporary design history, helping them to contextualise their practical activities and position their own practical endeavours. The researchers saw practice in context as being ideally suited to students using Web 2.0 tools to explore and document key historical and current influences on their field of study. Students used their first-year blog, accessed via the netbook, to upload written reflections, photos (geotagged via the smartphone), video reflections, interviews, web links and other original material captured using their smartphones. Thus, the focus was upon shared student-generated content and critiques by their peers and lecturers.

Students were required to undertake a written assignment in the form of an online blog that further elaborated on the weekly introductions to the history of contemporary product and furniture design. Building on their existing Vox blog, the students created a 'PIC Group'. This blog ran throughout the course, with students posting to their blog at least weekly, collating information about the people, movements, companies etc. covered in the weekly lectures, and writing up their self-directed research. Students were asked to comment regularly on one another's Vox blog posts – providing critique, feedback and links to appropriate resources. Unfortunately, the Vox e-portfolio site closed in mid-2010, leading to a migration to WordPress as the core student e-portfolio system.

Semester 2: Practice in context, assignment 2

The second assignment was built upon the processes and affordances of mobile Web 2.0 that students built up during the first PIC2 assignment. The assignment focused upon student-generated content, additionally using Web 2.0 tools to present to the rest of the class and the course lecturers. Students were required to create a chronological timeline (design-line) that identified and discussed key moments in design through products, craft objects, fashion, cars, architecture, exhibitions, literature, music, politics, war, graphics, manifestos, design schools, etc. Their design-line had to be visual as well as text-rich. A clear use of graphical communication was required. Student-sourced quotes from designers, industrialists and politicians were added to the design-line.

Final submissions could take the form of a booklet or poster hosted on Google Docs or some other form of digital product, such as a Picasa or Flickr annotated slideshow from an online web album with geotagged data, descriptions and mobile QR codes for sharing URL links via smartphones. Thus, the 2010 projects solidified the integration of mobile Web 2.0 tools within the foundational first year of the BPD course for student-generated e-portfolios, allowing students to capture and share their learning experiences in virtually any context, including formal and informal, using their mobile devices.

2011 and beyond

Building on the successful integration of mobile Web 2.0 tools within the course, the researchers explored the transferability of this approach across multiple learning contexts. Consequently, a group of like-minded lecturers from around the world were recruited to form a COP to explore the potential of mobile Web 2.0 tools for international collaboration between groups of students (Cochrane et al., 2011).

This international collaboration (the Icollab11 international project) aimed to enhance student teamwork by requiring students to utilise the communication and collaboration affordances of smartphones (for example Qik mobile video-streaming and Twitter) as they formed international teams and negotiated learning outcomes and team projects with the lecturers in all four countries creating a virtual cultural exchange experience. Students explored the roles of digital identities in online communities.

The project focused upon each student group forming a COP and sharing their own mobile-generated content as appropriate for their context and garnering feedback from the other international groups. Thus, each local physical COP was also augmented by a virtual COP made up of the participants from all four countries. Each lecturer brought unique m-learning experience and expertise to the collaborative project. The project involved exploring the use of:

- international collaboration;
- Google Docs for collaborative project mapping and management by the lecturers;
- Twitter for communication and sharing of ideas;
- student blogs for recording project progress and peer commenting;
- mobile student-generated media;

- remote presentation and peer critique of student-team projects;
- student-generated augmented reality;
- collaborative teaching.

The artefacts produced as reified learning objects by each student team COP became boundary objects that the lecturers used to broker the international virtual COP between the five local physical COPs. The goal was that this international project would facilitate pedagogical shifts from teacher-directed learning towards heutagogy or student-directed and negotiated learning contexts within the context of each of the participating countries. This project thus explored the potential of mobile Web 2.0 tools to enable and enrich student collaboration across international boundaries.

Mobile Web 2.0 framework: key aspects

Here follow some key aspects related to the use of a mobile Web 2.0 design framework to bridge multiple learning contexts, including formal and informal, and reconceptualise teaching and learning.

Critical success factors

Six critical success factors where identified across the 30 m-learning projects from 2006 to 2009 (Cochrane, 2010a), informed by the chosen social learning frameworks. These were identified as common critical issues when attempting to create new social constructivist learning environments in the variety of contexts explored by the researcher. Data analysis to reveal common themes was taken from pre-project surveys of participants, regular blog posts throughout the projects by the participants, post-project surveys of the participants, and focus groups from each project. The identified critical success factors were used to guide the design of the evolving m-learning projects, including the BPD m-learning projects (Cochrane, 2010b). The six identified critical success factors were:

- the pedagogical integration of the technology into the course and assessment;
- lecturer modelling of the pedagogical use of the tools;
- creating a supportive learning community;
- appropriate choice of mobile devices and Web 2.0 social software;
- technological and pedagogical support;
- creating sustained interaction that facilitates the development of ontological shifts, both for the lecturers and the students.

These critical success factors are informed by the chosen theoretical learning frameworks, and led to the development of a mobile Web 2.0 framework. These are explored in more depth in previous publications by the researchers (Cochrane, 2010a, 2012; Cochrane and Bateman, 2011a).

Scaffolding the PAH continuum

The concept of bridging the PAH continuum provided a measure of how much pedagogical transformation was achieved by each m-learning project, as the goal of each project was to develop a move from teacher-directed pedagogy towards student-directed heutagogy (Cochrane and Bateman, 2010b; Cochrane et al., 2009b). However, this transformation is a process, often involving a significant reconceptualisation of teaching and learning (or an ontological shift) by the participants. This process requires both pedagogical and technical scaffolding, and was achieved by the establishment of a collaborative partnership between the researcher as an academic adviser and technology steward, and the course lecturers and students.

The problem with technology-focused interventions in education tends to be their short-term nature as they typically involve an early-adopter educator working with final-year students. This leaves little scope for building on these projects or any significant sustained engagement for students. To move beyond such scenarios, our framework takes a staged and scaffolded approach to technology integration across the entire length of a programme, allowing significant sustained engagement for the participants, and making it possible to build on the appropriation of the technology and pedagogical change over multiple years, as illustrated in Table 3.1.

Facilitating ontological shifts via communities of practice

Each m-learning project was treated as the establishment of a unique learning community, modelled as a COP. The projects therefore put the most effort into nurturing and developing these COPs rather than the development of mobile-formatted content delivery or developing mobile platforms or applications (Cochrane, 2007; Cochrane and Narayan, 2011).

Framing the m-learning projects around the establishment of COPs facilitated regular and sustained engagement of the participants. Peer and expert modelling of the educational use of mobile Web 2.0 tools within these COPs helped the participants to reconceptualise their roles within a social-constructivist learning paradigm.

Table 3.1 Staging and scaffolding the PAH continuum

Stage	Learning context	M-learning project focus	Course timeframe	PAH alignment
Level 1	Social collaboration with peers and lecturer Student-generated content	Establishment of student-generated e-portfolios using student-owned mobile devices, learning management system and basic Web 2.0 sites	One-year certificate programmes, or first year of longer programmes	Pedagogy (lecturer directed)
Level 2	Social collaboration with peers and 'authentic environments' Context aware	Establishing the unique communication and collaboration affordances of student-owned mobile devices	Second year of two-year or longer programmes	From pedagogy to andragogy (students become the content creators)
Level 3	Context independent, bridging formal and informal learning Student-generated contexts	Student-negotiated team projects exploring (for example) mobile augmented reality using student-owned mobile devices	Third year of programme	From andragogy to heutagogy (students become independent learners)

Establishing a design framework

The design framework developed for each of the m-learning projects is shown in Table 3.2. This framework was developed iteratively over the life of the research, which began in 2006 with two test projects that informed the practical implementation of the subsequent projects from 2007 to 2011 (Cochrane, 2011; Cochrane and Bateman, 2011a).

The framework table format is based loosely on that suggested by Sharples et al. (2009). The emphasis is upon starting with desired learning practices then choosing appropriate technologies to manage and support these practices. This implies that the starting point of the design process is the learning practice and chosen pedagogical framework, which then informs the appropriate choice of mediating technologies.

Within the researcher's framework presented in Table 3.2, the elements of assessment, and pedagogical and technological support are explicitly dealt with within the lecturer pre-project COP and in the implementation stage of the project within the student and lecturer collaborative COP.

Mapping mobile Web 2.0 affordances

The m-learning projects explored how a mix of mobile Web 2.0 tools could enhance the students' learning throughout their whole course, and in particular how these tools could facilitate social constructivist learning environments, including student-generated content and student-generated contexts (Cochrane and Bateman, 2010a).

Table 3.3 provides an example of a mobile Web 2.0 tool and the pedagogical alignment with social constructivist activities and outcomes, building upon the mobile Web 2.0 framework described above.

Table 3.2 M-learning project design framework

Learning practice	**Mediating circumstances**		
Social constructivism	Context	Technology	Agent
Lecturer community of practice	Lecturer professional development, pedagogical brainstorming	Face to face, scaffolded using learning management system, smartphone, Web 2.0 services	Lecturers as peers, with researcher as technology steward
Student and lecturer community of practice	Pedagogical integration and technical support	Face to face, scaffolded using learning management system, smartphone, Web 2.0 services	Students as peers, lecturer as guide and pedagogical modeller, with the researcher as technology steward
Collaboration	Group projects	Social networking, collaborative documents	Google Docs, student peers
Sharing	Peer commenting and critique	Web 2.0 media sites, e-portfolio creation	RSS, student peers, lecturer
Student content creation	Student individual and group projects	Smartphone with camera and microphone, content uploaded to Web 2.0 sites	Student and peers
Reflective	Journal of learning and processes, recording critical incidents	Web 2.0 hosted blog	Personal appropriation, formative feedback from lecturer
Learning context bridging	Linking formal and informal learning	Smartphone used as communications tool and for capturing content	Student interacting with context, peers and lecturers

Table 3.3 Mapping mobile Web 2.0 affordances to social constructivist design

Learning practice	Mediating circumstances		
Social constructivism	Context	Technology	Agent
Group projects and team communication	Student collaboration across global boundaries, including questioning, commenting, polling and sharing of student-generated content via web links	Twitter via smartphone with embedded links to student-generated blog posts, YouTube videos, and polling via Polleverywhere	Modelled by expert lecturers, and appropriated by student peers

The researchers worked with lecturers to explore how mobile Web 2.0 tools could be integrated into new course activities and assessments that leveraged social constructivist pedagogy. This often involved the use of Google Docs for collaborative editing and brainstorming of course outlines and aligned assessment activities.

Discussion

The impact of the developed mobile Web 2.0 implementation framework described in this chapter is clearly seen in the changes implemented in the BPD course at Unitec from 2006 to 2010. Beginning with initial explorations of mobile blogging in the third year of the course in 2006, the course now features student-generated e-portfolios as a core assessment, formative feedback and a tool for collaboration.

The use of mobile devices (specifically smartphones) has enabled students to capture, record, share, curate and categorise ideas, processes and collaborative activities in and beyond the classical studio environment. Bridging formal and informal learning experiences has become explicit and authentic via mobile Web 2.0 tools.

The integration of lecturer and student COPs as a core support structure within the course has led to sustained collaboration between the course lecturers and the educational researcher as the technology steward. An outcome of this collaborative partnership has been not only pedagogical change but also significant practice-based research outputs that have been used to broker the concepts and approach to other lecturers in other learning contexts, including an international COP established in 2011.

While the commitment and input of the participants have been intense, the outcomes have been rich. The results have been significantly better than the previous approach of scaffolding educational technology integration via a short series of introductory workshops. Making project decisions explicitly founded upon a chosen social learning theory and associated learning frameworks has guided the implementation and development of the projects. The developed mobile Web 2.0 framework effectively satisfies Reeves' (2005) call for a new research methodology and new support strategies for educational technology integration. As Reeves (2009: 2) adds: 'design research requires intensive, on-going collaboration among researchers and practitioners to design and refine prototype e-learning environments tailored to the unique contexts in which they will be used'.

Thus, the research has resulted in the development of a transferable design framework and implementation strategy for mobile Web 2.0 in tertiary education that is sufficiently flexible to allow for new and unique contexts (Cochrane, 2010c). The implementation strategy matches the unique affordances of mobile Web 2.0 with social constructivist pedagogies, minimising the technical support needed for the participants, while maximising transferability.

The implementation strategy places the emphasis upon lecturer professional development and student scaffolding with the goal of transforming pedagogy rather than the development and programming of technically complex mobile software applications. The research developed explicit staging of the integration of mobile Web 2.0 within courses and scaffolding and nurturing of the required ontological shifts in pedagogical transformation via a structured and sustained intentional community of practice model over the length of the course, resulting in positive pedagogical change for both the lecturers and the students.

Coupled with an action research methodology, this ensured that each new m-learning project did not have to reinvent the wheel, but built upon the lessons learnt from previous projects. The range of learning contexts covered by the 2006–11 m-learning projects demonstrates the suitability of the implementation strategy for mainstream adoption.

Conclusion

This chapter outlines a design framework for implementing mobile Web 2.0 within higher education with the goal of transforming pedagogy from lecturer-directed pedagogy to student-directed heutagogy. This mobile Web 2.0 framework enables the design of student-generated content and learning contexts.

The BPD m-learning projects conducted between 2006 and 2010 illustrate how mobile Web 2.0 can be used as a catalyst to enable student-generated content and student-generated learning contexts in and beyond the classroom. Modelling lecturer professional development and class interactions as intentional COPs that are collaborative partnerships between the participants and educational technology researchers as technology stewards was found to provide a powerful support structure for pedagogical change and is a key component of the framework.

The framework is based upon social learning theory to guide implementation strategies and match the unique collaborative and communicative affordances of mobile Web 2.0 tools, ensuring that appropriate and transformational choices are made.

In summary, the mobile Web 2.0 framework can be used to support the design of m-learning projects that bridge learning contexts, including formal and informal (for example, in the creation of student-generated e-portfolios), physical and virtual (for example, the nomadic studio), situating learning within authentic collaborative student experiences (for example, the SHaC09 project and the Icollab11 international project).

Bibliography

Belshaw, D. (2010) *Mobile and Wireless Technologies Review*, Newcastle upon Tyne: JISC infoNet.

Brown, T. (2005) 'Beyond constructivism: exploring future learning paradigms', *Education Today* (2): 1–11.

Brown, T. (2006) 'Beyond constructivism: navigationism in the knowledge era', *On the Horizon* 14(3): 108–20.

Bryant, T. (2006) 'Social software in academia', *Educause Quarterly* 2: 61–4.

Chi, M. and Hausmann, R. (2003) 'Do radical discoveries require ontological shifts?' in L. Shavinina and R. Sternberg (eds),

International Handbook on Innovation, Vol. 3, New York: Elsevier Science, pp. 430–44.

Cochrane, T. (2006) 'Learning with wireless mobile devices and social software', *Proceedings of 23rd ASCILITE Conference: Who's Learning? Whose Technology?* Sydney, 3–6 December.

Cochrane, T. (2007) 'Moving mobile mainstream: using communities of practice to develop educational technology literacy in tertiary academics', *Proceedings of MLearn 2007 – Making the Connections 6th International Conference on Mobile Learning*, Melbourne, 16–19 October.

Cochrane, T. (2010a) 'Exploring mobile learning success factors', *ALT-J, Research in Learning Technology* 18(2): 133–48.

Cochrane, T. (2010b) 'An mlearning journey: mobile Web 2.0 critical success factors', paper presented at the MLearn 2010: 9th International Conference on Mobile Learning, Valletta, 26–30 October.

Cochrane, T. (2010c) 'Beyond the Yellow Brick Road: Mobile Web 2.0 informing a new institutional e-learning strategy', *ALT-J, Research in Learning Technology* 18(3): 221–31.

Cochrane, T. (2011) 'Exploring the pedagogical affordances of mobile Web 2.0', in L. Chao (ed.), *Open Source Mobile Learning: Mobile Linux Applications*, Hershey, PA: Information Science Reference, pp. 196–215.

Cochrane, T. (2012) 'An mlearning journey: mobile Web 2.0 critical success factors', *International Journal of Handheld Computing Research (IJHCR)* 3(2): 44–57.

Cochrane, T. and Bateman, R. (2010a) 'Smartphones give you wings: pedagogical affordances of mobile Web 2.0', *Australasian Journal of Educational Technology* 26(1): 1–14.

Cochrane, T. and Bateman, R. (2010b) 'Pedagogy 2.0: Invigorating a paper based design history course', *Proceedings of the EDULEARN10 the 2nd International Conference on Education and New Learning Technologies*, Barcelona, 5–7 July.

Cochrane, T. and Bateman, R. (2011a) 'Transforming pedagogy using mobile Web 2.0', in D. Parsons (ed.), *Combining E-Learning and M-Learning: New Applications of Blended Educational Resources*, Hershey, PA: IGI Global, pp. 281–307.

Cochrane, T. and Bateman, R. (2011b) 'Strategies for mlearning integration: evaluating a case study of staging and scaffolding mlearning integration across a three-year bachelor's degree', *Waikato Journal of Education (WJE)* 16(1): 109–24.

Cochrane, T. and Kligyte, G. (2007) 'Dummies2Delight: using communities of practice to develop educational technology literacy in tertiary academics', *Proceedings of the JISC Online Conference: Innovating eLearning*, 11–14 June.

Cochrane, T. and Narayan, V. (2011) 'DeFrosting professional development: reconceptualising teaching using social learning technologies', in D. Hawkridge, K. Ng and S. Verjans (eds), *Proceedings of ALT-C 2011 – Thriving in a Colder and More Challenging Climate: The 18th International Conference of the Association for Learning Technology*, 6–8 September, University of Leeds: ALT Association for Learning Technology, pp. 158–69.

Cochrane, T., Bateman, R., Buchem, I., Camacho, M., Gordon, A., Keegan, H., et al. (2011) 'Mlearning 2.0: fostering international collaboration', in I. Candel Torres, L. Gomez Chova and A. Lopez Martinez (eds), *ICERI2011: 4th International Conference of Education, Research and Innovations*, 14–16 November, Madrid: IATED, pp. 42–51.

Cochrane, T., Bateman, R., Cliffin, P., Gardner, J., Henderson, I. and Holloway, S. (2009a) 'Mobilizing learning: mobile Web 2.0 scenarios in tertiary education', *Proceedings of the EDULEARN09 the 1st International Conference on Education and New Learning Technologies*, Barcelona, 6–8 July.

Cochrane, T., Bateman, R. and Flitta, I. (2009b) 'Integrating mobile Web 2.0 within tertiary education', in A. Méndez-Vilas, A. Solano Martin, J. A. Mesa Gonzalez and J. Mesa Gonzalez (eds), *Research, Reflections and Innovations in Integrating ICT in Education*, Vol. 3, Badajoz: FORMATEX, pp. 1348–52.

Cook, J. (2010) 'Mobile phones as mediating tools within augmented contexts for development', *International Journal of Mobile and Blended Learning* 2(3): 1–12.

Garnett, F. (2010) 'Heutagogy and the craft of teaching', *The Heutagogic Archives*. Available at: *http://heutagogicarchive.wordpress.com/2010/11/18/heutagogy-the-craft-of-teaching/#more-340* (accessed 7 June 2009).

Hercules, F. (2001) 'Dyslexia in art and design higher education: Hidden strengths, hidden weaknesses', *Proceedings of the 5th British Dyslexia Association International Conference, DYSLEXIA: At the Dawn of the New Century*, York, 18–21 April.

Herrington, A. and Herrington, J. (eds) (2006) *Authentic Learning Environments in Higher Education*, Hershey, PA: Information Science Publishing.

Herrington, A. and Herrington, J. (2007) 'Authentic mobile learning in higher education', in P. L. Jeffery (ed.), *AARE 2007 International Educational Research Conference*, Fremantle: AARE.

Herrington, J. and Oliver, R. (2000) 'An instructional design framework for authentic learning environments. *Educational Technology Research and Development* 48(3): 23–48.

Herrington, J., Herrington, A., Mantei, J., Olney, I. and Ferry, B. (eds) (2009) *New Technologies, New Pedagogies: Mobile Learning in Higher Education*, Wollongong: University of Wollongong.

Kukulska-Hulme, A. (2010) 'Mobile learning as a catalyst for change', *Open Learning: The Journal of Open and Distance Learning* 25(3): 181–5.

Langelier, L. (2005) *Working, Learning and Collaborating in a Network: Guide to the Implementation and Leadership of Intentional Communities of Practice*, Quebec City: CEFIRO (Recherche et Études de cas collection).

Lave, J. and Wenger, E. (1991) *Situated Learning: Legitimate Peripheral Participation*, Cambridge: Cambridge University Press.

Luckin, R., Clark, W., Garnett, F., Whitworth, A., Akass, J., Cook, J., et al. (2010) 'Learner-generated contexts: a framework to support the effective use of technology for learning', in M. Lee and C. McLoughlin (eds), *Web 2.0-Based E-Learning: Applying Social Informatics for Tertiary Teaching*, Hershey, PA: IGI Global, pp. 70–84.

McLoughlin, C. and Lee, M. (2007) 'Social software and participatory learning: Pedagogical choices with technology affordances in the Web 2.0 era', in R. J. Atkinson, C. McBeath, S. K. A. Soong and C. Cheers (eds), *Proceedings of Ascilite 2007, ICT: Providing Choices for Learners and Learning*, Centre for Educational Development, Nanyang Technological University, 3–5 December, Singapore: Ascilite, pp. 664–75.

McLoughlin, C. and Lee, M. (2008) 'Future learning landscapes: transforming pedagogy through social software', *Innovate: Journal of Online Education* 4(5): 7–12.

McLoughlin, C. and Lee, M. (2010) 'Pedagogy 2.0: Critical challenges and responses to Web 2.0 and social software in tertiary teaching', in M. Lee and C. McLoughlin (eds), *Web 2.0-Based E-Learning: Applying Social Informatics for Tertiary Teaching*, Hershey, PA: IGI Global, pp. 46–69.

Pachler, N., Bachmair, B. and Cook, J. (2010) *Mobile Learning: Structures, Agency, Practices*, London: Springer.

Reeves, T. (2005) 'No significant differences revisited: A historical perspective on the research informing contemporary online learning', in G. Kearsley (ed.), *Online Learning: Personal Reflections on the Transformation of Education*, Englewood Cliffs, NJ: Educational Technology Publications, pp. 299–308.

Reeves, T. (2009, 16–18 March) 'The application of "design research" to e-learning', *Proceedings of the First International Conference for e-Learning and Distance Learning*, available at: *www.eli.elc.edu.sa/2009/content/Reeves[research].pdf* (accessed 18 December 2010).

Sharples, M., Crook, C., Jones, I., Kay, D., Chowcat, I., Balmer, K., et al. (2009) *CAPITAL Year One Final Report*, Nottingham: University of Nottingham.

Siemens, G. (2004) 'Connectivism: a learning theory for the digital age', *eLearnspace*, available at: *http://www.elearnspace.org/Articles/connectivism.htm* (accessed 9 October 2010).

Swantz, M. L. (2008) 'Participatory action research as practice', in P. Reason and H. Bradbury (eds), *The SAGE Handbook of Action Research: Participative Inquiry and Practice* (2nd edn), London: Sage Publications, pp. 31–48.

Vygotsky, L. (1978) *Mind in Society*, Cambridge, MA: Harvard University Press.

Wenger, E., White, N. and Smith, J. (2009) *Digital Habitats: Stewarding Technology for Communities*, Portland, OR: CPsquare.

Wenger, E., White, N., Smith, J. and Rowe, K. (2005) 'Technology for communities', in L. Langelier (ed.), *Working, Learning and Collaborating in a Network: Guide to the Implementation and Leadership of Intentional Communities of Practice*, Quebec City: CEFIRO, pp. 71–94.

4

Facing up to it: blending formal and informal learning opportunities in higher education contexts

Julie Willems and Debra Bateman

Abstract: Through an investigation of social networks, with a particular focus on Facebook, this chapter explores how Web 2.0 is blurring formal and informal teaching and learning opportunities in higher education. Using a knowledge-sharing framework brought about by the intersection of two continua (formal and informal learning opportunities; and student-directed and teacher-centred learning), and using a Facebook case study exemplar for each of the four quadrants of the framework, the chapter explores institutional (or teacher)-driven informal teaching and learning; student (or member)-driven informal teaching and learning; student (or member)-driven formal learning; and institutional (or teacher)-driven formal learning. The chapter concludes that an understanding of the potentials and pitfalls of each of these four environments is necessary in an era of rapid transformations in higher education.

Keywords: social networks, Facebook, student-directed learning, blended learning, higher education

Introduction

The teaching and learning landscape of academia is being transformed. Thanks to Web 2.0, social networks – often sites of informal personal communications – are swiftly becoming part of the learning journey within higher education (Willems and Bateman, 2011), encroaching upon traditional institutionally-driven or teacher-centred teaching and learning practices. Accompanying this transformation, the line between formal and informal learning is becoming increasingly blurred.

This chapter explores the landscape of formal and informal learning opportunities in higher education that are taking place via Facebook, a major social network. Adopting the four-quadrant knowledge-sharing framework created by the intersection of the twin axes of formal and informal learning opportunities, and student-directed and teacher-centred learning (Stuckey and Arkell, 2005), and using an auto-ethnographic and ethnographic approach, the chapter explores each quadrant through four representative case study exemplars from Facebook. These four quadrants are: institutional (or teacher)-driven informal teaching and learning; student (or member)-driven informal teaching and learning; student (or member)-driven formal learning; and institutional (or teacher)-driven formal learning. The Facebook exemplar for each quadrant serves to highlight the blurring of teaching and learning practices in higher education. The chapter concludes that an understanding of the potentials and pitfalls of each of these environments is necessary in an era of rapid transformations in higher education.

The blurring of formal and informal learning

Much of what is considered as taking place in higher education contexts relates to formal learning. Formal learning has different meanings depending on the context. For example, formal learning relates to the dominant education systems in any country (Hodkinson et al., 2003) and is governed by rules relating to its creation, operation and governance. It can be considered a form of high culture in that not all society members necessarily have access to it. Formal learning can also mean the site of structured learning under the auspices of an educational body which is premised on a set curriculum, a well-defined learning event, and the dedication of time towards the learning experience (Stone, 2009).

Formal learning is often associated with didactic, teacher-centric pedagogies and approaches designed to efficiently distribute, or 'push' (Hagel, 2005), the necessary and desirable resources or learning materials to the consumer (student). Stuckey and Arkell (2005) refer to these as 'institutionally-driven' approaches. Richardson (2006) suggests that in 'push' educational models, students are often considered as passive receivers of information. Grinnell (2009) concurs, arguing that this 'push' model of instructional delivery is 'similar to the prevailing information-transfer paradigm where teachers transfer "knowledge" to often passive, unengaged students'. Moreover, as Arif et al. (2005) argue, 'push' pedagogical models tend to promote 'one-size-fits all' approaches to teaching and learning which is often aligned with instructivist pedagogy.

Informal learning opportunities, on the other hand, are emergent, student-driven practices which rely on shared norms rather than rules. Facer (2004: 1) defines informal learning as 'the sort of learning that goes on as part of our normal day-to-day activities when we don't even think we are learning'. Siemens (2004) argues that informal learning is simultaneously the most dynamic and versatile, but least recognised, form of learning. He argues that it is based on information foraging, observing, seeking help, asking questions, and by trial-and-error. Indeed Cross (2003) argues that informal learning constitutes 80 per cent of what we learn in the workplace, while formal learning accounts for only 10–20 per cent.

Informal learning – also known as 'non-formal learning' (Hodkinson et al., 2003) – is often differentiated from formal learning because it generally does not involve the presentation of material for assessment, or the completion of exercises for feedback. Informal learning is often learner-driven, and encourages exploration and play, and the development of skills. Informal learning is commonly described as encompassing 'pull' approaches to teaching and learning (Hagel, 2005), relating to the demand-style practices of the consumer (student). In 'pull' approaches, the consumer demands, requests or creates the particular products or information that they require for their particular need. In educational contexts, these 'pull' approaches are often associated with student-centred (or 'member-driven') learning in which students are active and networked co-creators of their learning through innovative practices tailored to their specific needs (Richardson, 2006), sourcing appropriate resources whenever the need arises (Hagel, 2005). Consequently, 'pull' approaches to teaching and learning are more aligned with constructivist and connectivist pedagogies. Informal learning is often construed in a less positive light than formal learning, and is often connected with popular culture. Yet Coffield (2000: 8) argues that 'informal learning should no longer be regarded as an inferior form of learning whose main purpose is to act as the precursor of formal learning; it needs to be seen as fundamental, necessary and valuable in its own right'.

As a consequence, formal and informal teaching and learning are often viewed as the antithesis of each other. Hodkinson et al. (2003: 313) write that within the education sectors, this superficial division: 'reads as if there are two separate paradigms – informal (learning through everyday embodied practices; horizontal knowledge; non-educational settings) and formal (acquisitional and individual learning; vertical or propositional knowledge)'. Indeed, in contrast to the perception of formal and informal learning as oppositional and clearly demarcated perspectives, Stuckey and Arkell (2006) argue instead that they fall along a continuum of learning opportunities with purely formal learning opportunities at one end, and purely informal learning at the other. Evans and Larri (2007) argue that the blending of both formal and informal learning are essential because this meets both the institution's need to provide knowledge and skills acquisition and assess these, while simultaneously catering for the student's need to have innovative learning environments in which collaboration and self-directed learning can take place.

Thus, the traditional 'push' approaches to education are being challenged by 'pull' approaches of educational content co-created by students. The social spaces of Web 2.0 help enable these possibilities. Grinnell (2009) writes that 'Web 2.0's applications and platforms allow us to build communities of learners in which knowledge is constructed through conversation and interaction between all participants'. Web 2.0 offers a participatory medium in which social networking sites – such as YouTube, Facebook, MySpace and Twitter – allow individuals to congregate virtually, to view, create and share information beyond the confines of being present in the same physical space. Social networks are defined as enabling 'communication among ever-widening circles of contacts [and] inviting convergence among the hitherto separate activities of e-mail, messaging, website creation, diaries, photo albums and music or video uploading and downloading' (Livingston, 2008: 395). Social software is defined as software that is supportive of group interaction (Shirky, 2003) and the ability of the software users to develop an online profile of themselves within this bounded system. The social network, therefore, is created by users choosing to share their online profile and view the connections made by them in return (Boyd and Ellison, 2007). McLoughlin and Lee (2007: 667) note of social networks that 'in these spaces, youth engage in informal learning, and creative, expressive forms of behaviour and identity seeking, while developing a range of digital literacies'. Facebook is a prime example of social networking in Web 2.0 that highlights the blurring of formal and informal learning spaces as it originated in 2004 as a means of informally connecting university college students (Ellison et al., 2006; Grossman, 2010).

In these virtual spaces of Web 2.0, individuals are simultaneously consumers and producers: a situation suggested four decades ago by McLuhan and Nevitt (1972), with Toffler (1980) later coining the term 'prosumer'. Grinnell (2009) explains that prosumers are participatory 'consumers and producers of information and knowledge [and] share the stage as equals' with their educational guides. Bruns (2008), however, suggests that the term 'prosumer' is no longer appropriate, as although it reflects the use of the mass media in teaching and learning, it does not offer adequate explanation of the self-motivated creative developer of new content through the social-sharing nature of Web 2.0. He argues for a new term –'produsers' – as students in the modern era are simultaneously users who are 'also able to be producers of the shared information collection' (Bruns, 2009).

A new model for the knowledge economy

The incorporation of social software and social networks to blend formal and informal learning in higher education is a relatively recent phenomenon. As McLoughlin and Lee note:

> In such a digital world, with high connectivity and ubiquitous, demand-driven learning, there is a need to expand our vision of pedagogy so that learners are active participants or co-producers rather than passive consumers of content, and so that learning is a participatory, social process supporting personal life goals and needs. (McLoughlin and Lee, 2007: 664)

This is the essence of 'pedagogy 2.0': the 'interdependence between ideas, individuals, communities and information networks, supported by technology which is about linking minds, communities and ideas, while promoting personalisation, collaboration and creativity leading to joint knowledge creation' (McLoughlin and Lee, 2007: 668). With this, new ways of teaching and learning in higher education are arising through Web 2.0, blurring the boundaries between formal and informal learning in the context of higher education.

This blurring has been represented diagrammatically through the knowledge-sharing infrastructure or framework of Stuckey and Arkell (2005). The framework is created through the intersection of twin axes. The first axis is a continuum of formal and informal learning through technologically-mediated possibilities in Web 2.0. The second axis is a continuum spanning instructivist (teacher-centred and institutionally-driven) through to student-centred (member-driven) learning practices.

The intersection of these twin axes creates four quadrants to help explain the range of knowledge-sharing options: institutional (or teacher)-driven informal teaching and learning; student (or member)-driven informal teaching and learning; student (or member)-driven formal learning; and institutional (or teacher)-driven formal learning.

In the upper-right quadrant of this knowledge-sharing framework (Figure 4.1), formal and institutionally-driven knowledge-sharing practices are 'characterised as part of a "culture of compliance" where the institution has significant control over the method and timing of engagement', while in the bottom-left quadrant 'an "enabling culture" is characterised by support for individually-driven initiatives, guided by norms not rules and emerging from a bottom-up process' (Stuckey and Arkell, 2005: 11).

Stuckey and Arkell (2005) further suggest that in any knowledge-sharing framework, programmes and projects

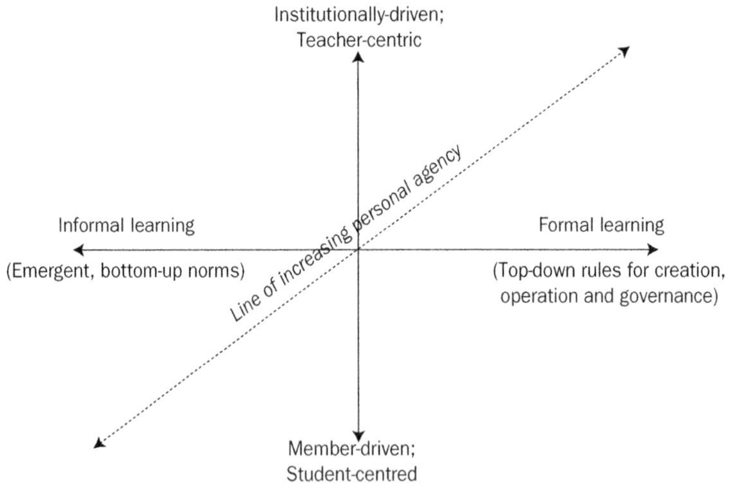

Figure 4.1 **Knowledge sharing framework**

Source: Based on Stuckey, B. and Arkell, R. (2005) *Development of an E-learning Knowledge Sharing Model*, Canberra, ACT: DEST, p. 14.

representing all four quadrants should be incorporated in order to ensure both a balanced and sustainable educational system. Evans and Larri (2007) prefer the term 'equilibrium' to describe this desired state of knowledge sharing.

Methodology

Any discussion or investigation that relates to the adoption of new technology also needs to consider the important constraints and challenges from a variety of sources, such as personal experiences (Johnson et al., 2011). Using both an ethnographic and auto-ethnographic methodological approach (Ellis and Bochner, 2000; Sparks, 2002), the personal experiences of educators using Facebook in higher education are explored as case study exemplars of the four quadrants of the knowledge-sharing framework (Stuckey and Arkell, 2005)

The data informing each exemplar have been drawn from focus group discussions, interviews and from observations, as well as anecdotal data in the case of auto-ethnography. In each case study, and in order to protect institutional, professional and personal identities, all identifiers have been significantly changed. For example, while the case describes a faculty of arts and education, the exemplar may have been drawn from a faculty of science. In many respects, the interchangeability of the data to construct these cases highlights a generalisability of practice and application across and within disciplines and pedagogies across higher education.

Facebook as an institutionally-driven informal learning environment

As an institutional entity, the faculty of arts and law at a large university has developed a series of pages and groups in Facebook to unite their students socially and academically. The pages and groups are all branded with the institution's logo and all linked to the formal institutional Facebook identity. The pages and groups are advertised on the faculty's webpage and are promoted within the institution's learning management system (LMS). Individual pages and groups are constructed by individual lecturers. Under the auspices of an institutional group, it is claimed that 'this site has been set up to enable discussions about the course in a less formal setting'.

One specific group has been created by a communications lecturer. The lecturer invites all students across a number of his subjects into this group, as well as maintaining students who have previously been enrolled with him. Within this group, the past and present students participate in spontaneous initiated discussions based upon wall posts, and 'present ideas, opinions and developments' through the uploading of media clips including film, Prezi presentations and digital stories. Using the fluidity of Web 2.0 environments, the lecturer and students use a range of tools, which are embedded within the wall post discussions or used as a post in their own right. Professional association activities, such as invitations to seminars from relevant employers and the possibility for students to seek employment, are promoted through the Facebook 'events' function. Through this informal environment, many networks 'naturally emerge as students and professionals connect with one another as an alumni group, where the commonality is the fact they are or have all been at the same Uni' (Debra Bateman, 2010, field notes). The lecturer says that:

> ... working in Facebook allows our community of practice to be much bigger than it is normally when we just work in the LMS. Only present students can work in there, and our discussions now are much richer because past graduates are in here as well. And, I also have colleagues from around the world and some of their students.

For this subject, the 'formal learning environment' where all learning materials are uploaded remains the LMS, in line with the institution's policy regarding 'web presence and online teaching and learning', and 'issues of intellectual property'. In the LMS, there are formal discussions directly related to the specific requirements of the subject, such as assessment and clarifications of content presented in lectures and tutorials. Within the LMS, there is interaction between students who are enrolled in both on-campus and off-campus modes of a specific subject. According to the lecturer, 'the LMS acts as a practice place for learning the theory or how things come to be and the Facebook groups act as a place to apply learning or test it in the real world'. In this way, the lecturer sees that he is 'helping students to build professional networks' and ensuring that they are able to identify how their learning is authentic.

Another advantage of these dual environments is that students can host materials that would be difficult or impossible to host in the LMS. The lecturer reports that in his institution it is difficult for students to share videos and images, other than as attachments. Using the social network they can embed their materials and then more seamlessly generate links from within their LMS. Similarly, the students are able to generate links from within the Facebook groups that others have posted, and to clarify them within the 'safety of their LMS'.

On the note of 'safety', the lecturer notes that one of the challenges of facilitating groups and pages on Facebook is his limited power to moderate what occurs within the virtual space. He suggests that while it is beneficial to have the dual environments, even when the group is created by 'invite-only', because it is branded as an institutional group, that he has a 'responsibility to make sure things are always done in a professional or collegial way'. He has experienced some conflicts with this commitment. In particular, while he has clear 'ground rules for engaging in his groups in the form of disclaimers', it has been difficult, on occasions to 'control the inputs from alumni' who are no longer viewing the course materials in the same ways.

Facebook as a member-driven informal learning environment

In a faculty of education, one subject requires that students undertake weekly tasks that are generated from the weekly lecture topics. After the introductory lecture, where the assessment task is introduced, each year's cohort of students has generated a Facebook group, in each of the four years of the subject's lifetime. The chair of the subject recalls:

> In the first lecture, I have told the students that I will not be providing any further information about the task outside of the weekly lectures. I do not put information on our online studies environment, nor do I reply to students who say they are unable to attend lectures for other reasons. In the case of students who have formal special consideration, I always provide them. But, all of the lectures are available online; they just have to take

> the time to listen to them. Each year, in this unit, a group of students have set up a Facebook group to avoid having to listen to the lectures or attend.

According to the chair of the subject, the students populate the Facebook group through a covert message system. They 'spread the word in other subject tutorials and through their own social networks'. Within these groups, students who have attended the lectures are asked to post the weekly tasks on the Facebook pages, for the 'other students to be able to access'. Then, they seek clarification among the students on what the tasks require, and some assurances about the ways that 'others have completed their tasks', as every week must be presented in a different way.

The Facebook group members upload files, such as help sheets on using different strategies to complete the weekly tasks as well as their weekly lecture notes. Often, individuals use the 'ask question' function to seek clarification on aspects of learning within a specific topic, but more commonly in addressing assessment criteria. To keep up with activities within the group, most users subscribe to all feeds, resulting in e-mail alerts regarding any further developments.

The chair of the subject sees this as extremely problematic, as students are not hearing the requirements of the tasks from the primary source, and may receive misinformation:

> In some ways, the Facebook group is a little like a group of students meeting in coffee shop somewhere to discuss what happens in their course. It's always happened and in many ways is what we want from them in their learning.

However, the chair of the subject recognises some nuances which magnify errors across more of the cohort. In the first instance, she claims that whereas in a coffee shop, there might only be a small number of students, in these Facebook groups, there can be hundreds of students and so the misinformation spreads further than in the coffee shop scenario. In the second instance, the chair feels concerned that the students are replacing the formal learning environment with the informal learning environment which is self-generated on Facebook. Because it is self-generated, none of the staff who teach in this subject are present in the environment and thus are unable to provide timely feedback on what is posted. Often, in the way that the group is populated, they are not aware of the group until late within a study period, by which time, considerable communications have already occurred within the group.

The chair also comments that the generation of the Facebook group creates a split between students enrolled in the subject. While at the beginning of the semesters, 'most of the students enrolled in the subject are members of the Facebook group, by the end of the semester only the students who are not attending lectures are present'. This, she explains, is because the students who are attending lectures 'become very resentful of the ones who do not go and are depending on them to do the work'. The chair reports that in one semester, it was clear which students were depending on the Facebook group, as they 'had received a fake task, and only the Facebook group submitted that one'.

Facebook as a member-driven formal learning environment

An institution has a thriving community of students who are enrolled in higher degree by research (HDR) courses. Historically, enrolment in these courses had been considered isolated, as students only met in face-to-face peer colloquia three times a year, and with their research supervisors, one-to-one, more frequently. The institution had set up a formal discussion board for the students to use; however, students rarely frequented the virtual space unless they had specific issues about their enrolment or about 'the mechanics of undertaking or writing up their research projects'.

After spending some time with one another at a 'chapter writing retreat', many of the students expressed a desire to have ongoing relationships and exchanged contact details, including making many reciprocal 'friend requests on Facebook'. As a result, a 'closed group', accessible only to students enrolled in this institution's HDR courses, evolved. Among the group, a small number of administrators were nominated, and the associated pages were created to reflect what were perceived as HDR student needs.

The pages of the group were titled 'About our research', 'About our methods', 'Times to talk' and 'Learning on the go'. 'About our research' provided each of the students with the opportunity to share information about themselves and their research project. Other students were encouraged to 'comment' on each others' posts with suggestions of further readings and to ask 'scholarly type questions'. More specifically, the 'About our methods' page offered students the chance to describe the methodological paradigm he or she was working within, and to align with others who were drawing upon similar readings, or clarifying their own research stances. The 'Times to talk' page offered regular opportunities for students to engage in peer forums, both online and face-to-face. These forums were student-generated, and occasionally invited external people, such as institutional scholars to participate within 'special interest groups'. The 'Learning on the go' page invited ongoing reflections from individuals at various stages of their candidatures, sharing their insights or experiences with others. All of the pages provided the facilities for students to upload links to external materials, for others to draw upon.

One of the administrators of the group, who has recently completed her studies, identified the Facebook group as 'the best ongoing support I had throughout the whole time I was enrolled'. It was in this group that this student 'could talk with others who were in a similar situation, where we're all researching different things, but we're all researching'. For her, it was in this space that she transitioned from 'just a student to being able to hold my own with the academics', as in this space, the students 'developed independence with others, and could develop our arguments because we knew each others' work'. She drew the parallel between the ways that these students worked within their formal community, and the ways that their academic supervisors were 'operating within their workplaces'.

Facebook as an institutionally-driven formal learning environment

The lecturer in the subject is celebrated as an innovative and exciting young lecturer in law. Within her subjects, she invites students to work creatively in the ways they interpret and represent their understandings of statutes. For example, one week, she asks the students to create a vodcast explaining any statute for a layman's audience. Often, students have commented that the institution's LMS limits the ways that they are able to respond to tasks 'in the same creative ways that we are able in external Web 2.0' based environments. In response, 'knowing that they [students] are already in other spaces like Facebook and Twitter', the lecturer considers it convenient and appropriate that the formal space which supports student learning for her units, is relocated to 'a space where they already are'.

The lecturer has created a formal group on Facebook for the purposes of the 'Introduction to Law' subject. Membership is limited to those enrolled within the specific semester, and a hyperlink redirects students from the institution's LMS into the Facebook login screen. The group site is constructed to reflect a similar format to that offered in the LMS, and contains links to weekly lecture notes, audio recordings of seminars, electronic readings, discussion boards and series of announcements. Students are encouraged to limit the availability of their posts to 'group members only', and not to link their personal status updates to the group's page.

The lecturer, having run this unit in the same way over two years, reflects that at the beginning of the semester 'not all students are on Facebook and some are not keen to be in there, but the majority are in the group site within moments that they know we'll use it'. She also claims that the traffic in her Facebook group appears to be more highly frequented than her site in the LMS, and that students interact more. However, she does caution that on Facebook 'there is more likely to be mixed social interaction among the academic work', which she assesses to be difficult at times. The lecturer also reflects that unlike the LMS where students are enrolled in subject groups through the student management system, she is continually managing requests and 'tags' from students who are prompted to join through Facebook's suggested friends and groups mechanisms.

On one occasion, the lecturer felt concerned that materials she posted in the Facebook group 'may have been available to others when a couple of students made links to things she had used'. This had occurred as a result of students sharing materials with 'public access'. The lecturer 'struggled with how unique [her] materials were when others could grab them from the internet' and commented that it had made her 'more thoughtful about what I put up there now. In the LMS, I don't even have to think about that'.

Discussion

What are the ramifications, then, of using social networks in teaching and learning in higher education: what lessons have been learnt? While personal presence in Facebook is already part of daily life for many students, institutional presence within Facebook is now on the increase. It is not uncommon to enter an institution's name within the search engine of Facebook and achieve numerous 'hits'. In many instances, the institutional name has been used by students. In some cases, students have generated peer-enrolment based groups, or used Facebook as a repository, or as a means for hosting materials for assessment tasks. In others, they use the environment as a venue to air grievances, challenge information as it has been presented, or critique the institution or people within the institution. At the same time, institutions are advertising their presence within social media as a way of 'keeping our students connected'.

Open source learning environments such as Facebook in the institutionally-driven informal learning environment enable a greater networking between, and beyond, particular student cohorts. Whereas in traditional virtual environments, folders have been set up for each discrete student cohort enrolled in a particular unit or course, within Web 2.0 environments, such as Facebook, there is the opportunity to build and connect student cohorts across time and space, as well as across courses and subjects. This is beneficial in terms of developing communities and alumni within higher education settings; however, at the same time, issues of intellectual property and control are problematic for institutions and individuals (Bateman and Willems, 2012). Further dilemmas associated with responsible and ethical behaviours online, while engaged in studies and notions of duty of care, also challenge the role of Web 2.0 and Facebook. Certainly, one challenge striking the authors is how 'informal' a learning environment can be when it is actually initiated, structured and bounded by a formal learning instructor, environment and organisation. We would argue that an environment initiated by a person associated with an institution carries that institution's duty of care, code of conduct and ethical guidelines for engagement into any learning environment or research relationship with which they engage, in the name of the organisation.

In most institutions of higher education, and more broadly across the educational sector, there is an increasing claim that in responding to the impetus of the knowledge economy, we need to be working in more 'multimodal' ways (see, for example, Gee, 2007; Kalantzis, 2007; Kress, 2009; New London Group, 2000; Yelland et al., 2008). An advantage of access to and use of these increasing Web 2.0 and open source spaces is that it challenges instructors, learners and administrators to expand the repertoire of digital and representation modes in the ways that knowledge can be represented, and more importantly critically engaged upon and/or with. Moreover, the use of informal and formal learning environments situated within open source and social media, enables co-learning and technology-infused interactions. In turn, the opportunities for deliverables by higher education institutions in delivering greater flexibility, accessibility and versatility are also magnified by the contrast of traditional physical lecture and tutorial modes of engagement and learning. On the other hand, however, the challenge for institutions to supply and support staff and students with sufficient access to hardware, mobile devices, download memory space and internet access/speed pushes an already pressed sector of educators to face further financial pressures, and demands a rethinking of the configuration of physical spaces (Mason and Bateman, 2011). It also challenges institutions to provide sufficient resources to enable just-in time and ahead-of-the-game forms of technological professional learning.

Conclusion

Web 2.0 and the use of social networks in teaching and learning practices in higher education are evolving at a rapid pace. Using Stuckey and Arkell's (2005) 'knowledge-sharing framework', this chapter has explored the blurring of formal and informal teaching and learning across four quadrants: institutional (or teacher)-driven informal teaching and learning; student (or member)-driven informal teaching and learning; student (or member)-driven formal learning; and institutional (or teacher)-driven formal learning. The chapter has examined these quadrants through case study exemplars from Facebook, although the insights shared here can apply to any social networks. Each exemplar has served to highlight the blurring of boundaries, the possibilities and opportunities, and also the pitfalls from a teaching and learning perspective. As Bateman and Willems (2012) have noted, the pitfalls of using informal social networks in higher education can lead to dangerous territory for student, teacher and institution alike and so care must be taken in setting parameters within the trend.

Whereas many pedagogical and technological approaches may be easily described within a framework of best practices which have been identified, tested and described over time, the uses of generative technologies require much greater adaptability. In the same ways as pedagogies within physical spaces can transcend disciplinary content, the use of virtual environments can also be employed in a range of ways that traverse higher education courses and disciplines. It is only through documented experience and reflective practices that new commonalities, challenges and affordances can be recognised, suffice to say that there must still be adequate clarities and guidelines to protect the interests of all stakeholders. For while guidelines for putting information 'out there' are, to a degree, already established through policies around intellectual property from an institutional perspective; branding protocols are governed through the institution's marketing department; policies on the ways in which teaching and learning are to be conducted are well developed across the sector, and nuanced within the specific employment and enrolment context; and protocols around teacher–student relationships and interactions are also guided; what are new, and must be further interrogated, are the amalgams of each of these as they interact in new amorphous places which are borderless and subject to varying jurisdiction. In some ways, this notion could be challenged through the parallels of higher education programmes, such as 'philosophy in the pub', where formal and informal learning encounters are situated external to the institution, and in a similar public environment to that of the public web. However, in the way that the pub is still subject to governance by the physical innkeepers and the police of a society, the shape-shifting and mobile genres of social networks and Web 2.0 tools demand greater clarity and articulation of expectation and reinforce notions of social and educational contracts and implicit trust, in lieu of the external overseers.

Thus, the landscape of teaching and learning in higher education is changing. While it is impossible to halt this progress, it is possible to be aware of the potentials and pitfalls and to make informed decisions about them. To do so is to work with the blurring of boundaries between formal and informal learning and not to work against them.

Bibliography

Arif, M., Smiley, F.M. and Kulonda, D.J. (2005) 'Business and education as push-pull processes: an alliance of philosophy and practice', *Education* 125(4): 602–14.

Bateman, D. and Willems, J. (2012) 'Facing off: Facebook and higher education', in C. Wankel and L. Wankel (eds), *Misbehaviour Online*, Bingley: Emerald Publishing.

Boyd, D. and Ellison, N. (2007) 'Social network sites: definition, history and scholarship', *Journal of Computer-Mediated Communication* 13(1), available at: *http://jcmc.indiana.edu/vol13/issue1/boyd.ellison.html* (accessed 6 September 2010).

Bruns, A. (2008) *Blogs, Wikipedia, Second Life, and Beyond: From Production to Produsage*, New York: Peter Lang.

Bruns, A. (2009) 'From prosumer to produser: understanding user-led content creation', paper presented at Transforming Audiences Conference 2009, London, 3–4 September, available at: *http://produsage.org/node/67* (accessed 13 November 2010).

Coffield, F. (2000) *The Necessity of Informal Learning*, Bristol: Policy Press.

Cross, J. (2003) 'Informal learning – the other 80%', available at: *http://www.internettime.com/Learning/The%20Other%2080%25.htm* (accessed 4 July 2006).

Ellis, C. and Bochner, A. P. (2000) 'Autoethnography, personal narrative, reflexivity: researcher as subject', in N. K. Denzin and Y. S. Lincoln (eds), *The Handbook of Qualitative Research* (2nd edn), Newbury Park, CA: Sage, pp. 733–68.

Ellison, N., Steinfeld, C. and Lampe, C. (2006) 'Spatially bounded online social networks and social capital: The role of Facebook', paper presented at the Annual Conference of the International Communication Association, Dresden, 19–23 June.

Evans, V. and Larri, L.J. (2007) 'Social software and its possible future uses', *Australian Flexible Learning Framework: Supporting e-learning opportunities (Part III)*, available at: *http://pre2009.flexiblelearning.net.au/flx/webdav/site/flxsite/shared/Research%20and%20Policy%20Advice/Learning_with_Social_Software_Exec_Summary.pdf* (accessed 21 July 2011).

Facer, K. (2004) 'Foreword', in J. Sefton-Green (ed.), *Report 7: Literature Review in Informal Learning with Technology Outside School*, available at: *http://archive.futurelab.org.uk/resources/documents/lit_reviews/Informal_Learning_Review.pdf* (accessed 21 July 2007).

Gee, J. (2007) 'Pleasure, learning, video games, and life: the projective stance', in M. Knobel and C. Lankshear (eds), *A New Literacies Sampler*, New York: Peter Lang, pp. 95–113.

Grinnell, C. (2009) 'Wagging the long tail: from push to pull to create the writing center 2.0', *Praxis* 6(2), available at: *http://projects.uwc.utexas.edu/praxis/?q=node/258* (accessed 2 March 2011).

Grossman, L. (2010) 'Person of the year 2010: Mark Zuckerberg', *Time Magazine*, available at: *http://www.time.com/time/specials/packages/article/0,28804,2036683_2037183_2037185,00.html* (accessed 16 June 2011).

Hagel, J. (2005) 'From push to pull', *Viewpoint*, available at: *http://www.johnhagel.com/view20051015.shtml* (accessed 19 November 2008).

Hodkinson, P., Colley, H. and Malcolm, J. (2003) 'The interrelationships between informal and formal learning', *Journal of Workplace Learning* 15(7/8): 313–18.

Johnson, L., Smith, R., Levine, A. and Haywood, K. (2011) 'The 2011 Horizon Report', available at: *http://net.educause.edu/ir/library/pdf/HR2011.pdf* (accessed 12 January 2011).

Kalantzis, M. (2007) 'Knowledge and learning in communities of practice: changing contexts for continuing education in the professions', in P. Kell and G. Vogl (eds), *Higher Education in the Asia Pacific: Challenges for the Future*, Newcastle: Cambridge Scholars Pub, pp. 103–29.

Kress, G. (2009) *Multimodality: A Social Semiotic Approach to Contemporary Communication*, London: Routledge.

Livingston, S. (2008) 'Taking risky opportunities in youthful content creation: teenagers' use of social networking sites for intimacy, privacy and self-expression', *New Media & Society* 10(3): 393–411.

Mason, C. and Bateman, D. (2011) 'Deakin University: Flexibility in spaces', paper presented at the 2nd Annual Learning Space Design Summit, Informa Higher Education Series, Melbourne, VIC, 5–6 December.

McLoughlin, C. and Lee, M. J. W. (2007) 'Social software and participatory learning: pedagogical choices with technology affordances in the Web 2.0 era', paper presented at ascilite 2007 – ICT: Providing Choices for Learners and Learning, Singapore, 2–5 September, available at: *http://www.ascilite.org.au/conferences/singapore07/procs/mcloughlin.pdf* (accessed 9 November 2009).

McLuhan, M. and Nevitt, B. (1972) *Take Today: The Executive as Dropout*, New York: Harcourt Brace Jovanovich.

New London Group (2000) 'A pedagogy of multiliteracies: designing social futures', in B. Cope and M. Kalantzis (eds), *Multiliteracies: Literacy Learning and the Design of Social Futures*, South Yarra, VIC: Macmillan, pp. 9–37.

Richardson, W. (2006) *Blogs, Wikis Podcasts, and Other Powerful Web Tools for Classrooms*, Thousand Oaks, CA: Corwin Press.

Siemens, G. (2004) 'Categories of e-learning', available at: *http://www.elearnspace.org/Articles/elearningcategories.htm* (accessed 22 March 2010).

Shirky, C. (2003) 'A group is its own worst enemy: Social structure in social software', available at: *http://www.shirky.com/writings/group_enemy.html* (accessed 2 May 2006).

Sparks, A. (2002) 'Autoethnography: self-indulgence or something more?' in A. P. Bochner and C. Ellis (eds), *Ethnographically Speaking: Autoethnography, Literature, and Aesthetics*, Walnut Creek, CA: Alta Mira Press, pp. 209–32.

Stone, T. (2009) *Blending Web 2.0 Technologies with Traditional Formal Learning: A Guide for CLOs and Training Managers*, Rochester, NY: Element K Corporation, available at: *http://www.training-partners.com/downloads/elearning/Blending%20Web%202[1].0%20-%20Rev%200409.pdf* (accessed 22 March 2009).

Stuckey, B. and Arkell, R. (2005) *Development of an E-Learning Knowledge Sharing Model*, Canberra, ACT: DEST, available at: *http://dspace.flexiblelearning.net.au/dspace/bitstream/10165/2783/1/Development_of_an_%20e-learning_knowledge_sharing_model.pdf* (accessed 16 June 2010).

Toffler, A. (1980) *The Third Wave*, New York: Bantam Books.

Willems, J. and Bateman, D. (2011) 'The potentials and pitfalls of social networking sites such as Facebook in higher education contexts', paper presented at ascilite 2011 – Changing Demands, Changing Directions, Hobart, 4–7 December.

Yelland, N., Cope, B. and Kalantzis, M. (2008) 'Learning by design: creating pedagogical frameworks for knowledge building in the twenty-first century', *Asia-Pacific Journal of Teacher Education* 36(3): 197–213.

Networked lives for learning: digital media and young people across formal and informal contexts

Solveig Roth and Ola Erstad

Abstract: Much educational research focuses on classroom learning and reveals the struggle with using digital media as part of school activities. Studies describe how formal and informal digital media practices differ. The interconnections between these two life-worlds experienced by young people present a challenge to researchers. To investigate such networks, the authors have used a learning ecology methodology to make explicit contexts found in physical and virtual spaces that provide opportunities for learning. Each context comprises 'funds of knowledge' or social and cultural resources. The authors asked: how can digital media support and enhance the interconnection between learning practices in and out of school? This study of three 15-year-old students in a multi-ethnic suburb in Oslo, Norway, illustrates how personal interest in digital media spurred by 'funds of knowledge' in informal contexts has aided the development of formal learning. Such practices enable students to use research-based methods and structured problem solving and to draw on interests and expertise from different contexts. The use of digital media has catalysed improved learning identity. Networked lives for learning represent the new platform where research needs to explore and understand the interconnections between formal and informal contexts as we approach learning in the twenty-first century.

Keywords: networked lives, funds of knowledge, personal interest, digital media, formal–informal learning

Introduction

Much educational research has focused on learning activities within the classroom (Sawyer, 2006). During the last decade, the influence of digital technologies on classroom activities has become a key area of educational research (Kozma, 2003). This research shows how teachers and students struggle to implement and define fruitful learning practices using digital media in different subjects and at different levels in school (Law et al., 2008). Further, critics often describe institutional practices as barriers to school development related to the integrated use of such media. Several critical voices have also found fault with strategies that push the implementation and use of digital media in schools (Selwyn, 2011). In contrast, research on young people's use of different media outside school shows how practices using digital media differ from practices inside school in both form and content. The literature has described leisure-time activities employing digital media as an alternative route to engagement and learning better adapted to twenty-first-century needs than traditional school learning (Gee, 2003; Ito et al., 2010).

Finding ways to understand the interconnections and networking between these two life-worlds experienced by young people presents a major challenge to educators. During the last decade, the research community has witnessed a growth in cross-discipline interest in understanding better how knowledge travels from one setting to another and how learners experience this in their everyday lives and practices – online and offline. We can see this in ways of rethinking key concepts such as 'context' (Edwards et al., 2009), 'trajectories' (Dreier, 2003) and 'identity' (Lemke, 2007; Wortham, 2005).

In exploring such issues, we believe the concept of *network* is important to understand social transformations. However, our conception of network does not follow the line of such social theorists as Manuel Castells et al. (2006), the most famous of the network theorists, nor does it follow the actor-network theory of Bruno Latour (2005) in ways of studying subjects and objects as social agents. Network for us implies, to a larger degree, studying the network of funds of knowledge that young people draw on in their learning trajectories between home, community and school. We focus mainly on their media practices and the role this has on networking mechanisms in their learning lives.

Our research question is thus: how can digital media support and enhance the interconnection between learning practices in and out of school? We want to highlight how digital media – everything from computers and mobile technologies to social media – work in the everyday learning lives of young people. The first step in the presentation of data outlines quantitative data showing access and use of digital media among Norwegian youth both inside and outside of school. The next and main part of data presentation derives from an ongoing project in eastern Oslo characterised by cultural diversity. We follow three 15-year-olds to understand how out-of-school interests and activities using digital media play a role in learning trajectories in school settings and how they draw on different funds of knowledge that play a role in the ways they position themselves as learners in networks online and offline.

Networked lives

During the last decade, the amount of time young people spend on digital media has grown steadily. For example, 2009 marked the first year that the average time spent with media by Norwegian youth (16–24-year-olds) exceeded nine hours per day (Erstad, 2010). This implies that most young people multitask, i.e. spend time with different media simultaneously: watching television, listening to music, using Facebook and sending messages via SMS.

The Nordic countries (Denmark, Sweden, Norway, Finland and Iceland) interest us partly because of the penetration of technologies in all aspects of society coupled with a high degree of access among young people. At the same time, these countries share a public and political awareness about the importance of digital literacy related to education and do not put strong emphasis on testing as do many other countries. Hence, we might very well call this region the land of the digital natives.

Access to computers, the internet and mobile phones with internet capabilities among Nordic youth is more than 90 per cent (ages 16–22) and in some areas up to 100 per cent. Thus, access is not an issue for most youth in these countries. During the last decade, the time spent on digital media among Nordic youth (age range 8–24 years) has grown steadily (Nordicom, 2009). For example, 73 per cent of all 8–18-year-olds use the internet daily (Norwegian Media Authority, 2008). In Norway, Synovate's annual study on media use among children and youth shows that on average, the time spent on media during a regular weekday has increased from 4 hours and 45 minutes in 1991 to 6 hours and 32 minutes in 1999 to 7 hours and 58 minutes in 2009 (Erstad, 2010).

Social networking sites have existed in the Nordic countries for a relatively long time. Some examples include

LunarStorm in Sweden, mPetreklanen in Norway and Skum.dk in Denmark. The development of these sites started around 1999–2000, with young people demonstrating increasing interest in social networking; in recent years, however, Facebook has taken over as the main social media site favoured by youth (Erstad et al., 2007).

The implementation of digital technologies within the Nordic education systems has progressed with strength from the beginning of the 1990s onward (Erstad and Quale, 2009), focusing mainly on the technology itself and on accessing computers and the internet in schools. Only in recent years have critiques, reflection and issues around production and creativity started to emerge in terms of the ways schools put digital media to use (Erstad and Gilje, 2008).

In studying networked lives, we need to go beyond issues of access and context-bounded use and look more closely at the everyday practices of young people and how digital media create different trajectories of learning for different individuals. In their US studies, Mimi Ito et al. (2010) describe this as 'media ecologies'. In their large-scale 'Digital Youth' project, they manage to document the broader social and cultural contours, as well as the overall diversity, in youth engagement with digital media. The concept of *ecology* is used strategically to highlight that:

> The everyday practices of youth, existing structural conditions, infrastructures of place, and technologies are all dynamically interrelated; the meanings, uses, functions, flows, and interconnections in young people's daily lives located in particular settings are also situated within young people's wider media ecologies ... Similarly we see adults' and kids' cultural worlds as dynamically co-constituted, as are different locations that youth navigate such as school, after-school, home, and online places. (Ito et al., 2010: 31)

In their findings, the authors draw out certain genres of participation, in what they describe as 'friendship-driven' and 'interest-driven' categories. Further, they identify different levels of commitment and intensity in new media practices. They interpret these genres of participation as 'intertwined with young people's practices, learning, and identity formation within these varied and dynamic media ecologies' (Ito et al., 2010: 31). However, we should be careful in emphasising differences between online and offline activities. As Mark Nunes (2006) has made explicit, we live in the intersection between the online and offline as part of our everyday practices. In exploring digital youth, it is also important not to get caught up in overly general conceptions (Buckingham and Willett, 2006). Youth vary greatly in their degrees of digital competence and technological interests. Sonia Livingstone's (2009) studies of the bedroom cultures of young people using digital media in the UK is an example of studies that raise critical questions about what we mean by 'digital youth' and the role of media in young people's everyday lives.

Learning lives

The phrase *learning lives* (Edwards et al., 2010) refers to the coherence between learning, identity and agency in the individual, framed by a biographical approach studying peoples' learning trajectories over the life course. Personal histories and future orientations go toward creating 'narratives of the self'; these selves become central to productive learning. In relation to a 'learning-lives' approach, the connection between learning and identity becomes important because it specifies how different learners engage in learning activities across settings (Hull and Greeno,

2006). Learning does not end when one leaves the school grounds at the end of the day.

Wortham (2005) describes how social identification and academic learning depend deeply on each other by closely following the 'learning identity' among a few students across an academic year. Packer and Goicochea (2000) describe this as *ontological learning*, the process through which individuals create and transform themselves as they interact with others, sign systems and the world. Such studies have significance because they give in-depth understanding to learners' involvement in different processes of growth and how school and everyday cultures influence their development as persons and learners (McLeod and Yates, 2006).

Challenging our conceptions of 'context' is important because it informs us in an analytical sense of the way we interpret and understand the interrelationship between people, their learning identities, and their circumstances at different times and in different places. Edwards et al. (2009) relate the discussion on context to the broader discourse of lifelong learning where context refers to an outcome of activity or is itself a set of practices. In emphasising the process of networking between people and environments, they use the term *contextualizing* rather than *context.* Practices are not bounded by context but emerge relationally and are polycontextual, i.e. have the potential to be realised in a range of strata and situations based on participation in multiple settings. Once one looks beyond the context of conventional situations for education and training – allowing the extension of learning contexts into the dimension of relationships between people, artefacts and variously defined others mediated through a range of social, organisational and technological factors – the limitations of much conventional pedagogy come into sharp focus (Edwards, 2009). This raises an important point concerning how contextualising and networking involve different types of learning and different contents and implies different purposes which might be variables in the values defined for them.

As a continuation of the above discussion of context and contextualisation, an interest has arisen in elaborating on the concepts of 'space' and 'place' related to learning and the use of digital technologies in the knowledge society (Bekerman et al., 2006; Leander and McKim, 2003). The internet in particular has created whole new ways of conceptualising learning spaces and how they relate to everyday life (Nunes, 2006). The key is to track meaning-making across contexts over time to describe individual learning trajectories. Leander et al. (2010) promote new ways of evaluating how learning takes place, in what they term as 'new mobilities'. In their review, Leander et al. (2010) show how projects within different disciplinary fields can enlighten us about ways of tracing spaces and places that engage individuals using different resources.

Adolescents often pursue learning opportunities inside and outside school once they become interested in a topic. 'A learning ecology' study conducted by Barron (2006) shows how interest-driven learning inside or outside of school can be connected in a networked life for learning and thus catalyse development.

A funds-of-knowledge study conducted by Gonzales et al. (2005) documents these funds of knowledge and literacy practices within immigrant households in the USA. The concept refers to the thesis that one has historically and culturally developed knowledge and competences necessary for household and individual purposes; social experience and labour give competence. Such sources of knowledge can also consist of peer groups, community and popular-culture-based areas and sources of knowledge (Moje et al., 2004). Here, 'funds of knowledge' means categories and contents found in everyday life and how this knowledge is grounded in the 'thick' social and cultural relations that make up a learning life (Moll et al. 1990).

The socio-cultural theory based on Vygotsky's (1978) ideas emphasises the social and culture-dependent nature of learning. Through participation, the individual learns to manage the tools of thinking and acting appropriate for the community (Lave and Wenger, 1991). Thus, learning represents a social process that reshapes itself continuously as individuals and communities act in various environments and spaces (Wenger, 1998). When studying networked lives in this context we investigate how young people connect funds of knowledge and digital media in everyday practices, creating trajectories that enhance learning.

Methodology and research context

Methodology

Our overall challenge in the ongoing 'learning lives' project[1] is to understand the coherence between learning, identity and agency of the individual framed by a biographical-narrative approach and studying learning trajectories over the life course. In addition, we characterise learning as the capacity to adapt to changing roles within different contexts (Holland et al., 2003).

We focus both on 'learners' in ethnographic biographical life-history narratives and 'learning' in the socio-cultural perspective (Sawyer, 2006). Therefore, this research has two different perspectives in mind: the informants' biographical narratives and learning trajectories within contextual framings. The terms *biographical methods* and *life-history research* cover a wide range of approaches with a common methodological starting point – the collection and analysis of biographical accounts. This includes a number of research traditions with distinct aims, fields and methods of enquiry, with a focus on temporal processes (Thomson, 2009).

To achieve this, a learning ecology methodology (Barron, 2004; 2006; Ito et al., 2010) – defined as the set of contexts found in physical or virtual spaces that provide opportunities for learning – has guided the data collection. Each context comprises configurations of activities, material recourses, relationships and the interactions that emerge from them. Thus, it can explore funds of knowledge as sources and areas for learning.

The framework draws on ecological, socio-cultural and activity-theory perspectives and builds on prior studies of out-of-school learning (Hull and Schultz, 2001). It takes into account the possibility of permeable boundaries and young people's tendencies to draw on multiple cultural forms facing their current needs. Finally, it recognises that learning can be intertwined with processes of identity authoring and, when necessary, secondary developmental processes can emerge from more distant learning events (Barron, 2006).

The framework raises questions about learning by identifying how adolescents, once interested, create learning opportunities or capitalise on those offered. It highlights the usefulness of documenting the multiplicity of pathways in learning contexts that learners might take and raises questions about the emergence of interest, specifically, how additional social, material or identity resources spark or inhibit and how processes of consideration of fit between a topic and one's sense of self or identity are implicated. It may provide some examples of moves that adolescents make to create new activity contexts for themselves that then support their learning (Barron, 2006).

Analytical perspectives

The topic investigates how digital media enhance the interconnection between learning practices inside and outside school. It has here a special focus on how digital media connect with funds of knowledge, allowing new kinds of agency in learning. Four main conjectures about the dynamics have guided the data collection: (1) examination of pathways of participation to provide an account of events, activities and processes that spark interest in learning; (2) knowledge grounded in the social and cultural relations that make up a person's everyday life; (3) categorisation of creation of learning opportunities; and (4) mapping changes in a person's learning ecology, including consequential transitions.

Portraits follow of three youths selected from a group of 14–15-year-old participants from the ongoing learning lives project and followed over two years from secondary to upper secondary school. The participants were recruited from two schools in Oslo. At each school one class of students was sampled. The students in the classes represented a variety of ethnic groups in the Grorud Valley. The majority of the students responded positively to our request for participants. As such, the 14 participants represented were selected randomly. The three youths described in this paper were chosen because they showed particular interest in using digital media and technologies.

A biographical-narrative approach allows us to chart a learning history including the meaning and attribution behind decision making and narratives of how learning activities unfold across time, resources and historical context (Bruner, 1994). The interviews provide us with portraits that to some degree 'recover the person' in our theorising about human development (Mishler, 1996).

The data were generated through participant observation, formal and informal open-ended interviewing and use of the informants as co-researchers (inspiration found in the learning ecology perspective). Observations were conducted at school in spring 2011, in leisure time and in home visits from autumn 2011 to spring 2012. The interviews presented here were undertaken with each informant from autumn 2010 to spring 2012. The goal is to provide illustrations of the phenomena rather than prove their ubiquity.

The analytical categories used for coding each portrait are 'digital media and interests as funds of knowledge', 'learning identity' and 'from informal to formal'. This first category contains descriptions of how personal interest emerges in relation to everyday knowledge: digital media, technologies, traditions and family/cultural history. The second category contains descriptions of how identity, learning and agency interplay and influence the positioning of the learner. The last category describes how the learner uses informal knowledge in a formal setting.

We do not want to generalise as the examples are based partly on retrospective accounts; they do show a common seeking out of resources or creating new activity contexts among the youth to learn how to use digital media. The cases represent generative cross-context pathways. We are interested in documenting cases where digital media and interest are connected to funds of knowledge.

The research context

The Grorud Valley (130,000 inhabitants), situated in the eastern part of Oslo (600,000 inhabitants), was largely inhabited in the 1950s and 1960s by ethnic Norwegian working-class families. It expanded gradually with industry and commercial centres surrounded by different local housing neighbourhoods. Since the 1960s, many refugee and immigrant groups have moved into the Grorud Valley, with different ethnic groups dominating in different periods (Nielsen, 2009). The valley has many minority languages, and many neighbourhoods have more than 35 per cent immigrants – some up to 90 per cent. Compared with other areas in Europe, the Grorud Valley is not a ghetto with visible poverty and particular social problems. Most of the inhabitants own their own apartments or houses. However, material and immaterial resources available for the youth differ greatly among different neighbourhoods (Bråthen et al., 2007). The ethnographic fieldwork presented comes from two different neighbourhoods.

Three portraits

Again, this study looks at three students living in the Grorud Valley. Hanne, an ethnic Norwegian girl, lives with her grandmother, step-grandfather and sister in an apartment. Her family settled here after leaving a small agricultural community in western Norway in the 1960s. Aisha and Ugur are second-generation immigrants born in Oslo. Aisha, with a Gambian background, lives with her uncle and her sister in a small apartment. Ugur, with a Turkish background, lives with his parents and siblings in an apartment. Aisha and Ugur speak Norwegian with friends and their native tongue at home.

Based on their parents' education levels and the bilingual ability of two of them, their cultural capital (cf. Bourdieu, 1986) differs from those who positively rely on the traditional transfer of academic skills (Krange and Øya, 2005). The participants confirm this in the sense that they received average grades in secondary school. Their transition to upper secondary school represented some change in school performance; Aisha performs well in her school's services and business programme. Hanne struggles with maths but does well in the media and communication programme. Ugur performs mostly as before and does well in maths and scientific project work in the academic programme. His parents feel positively about the use of computers for enhanced education and learning.

The three have many leisure time activities. Aisha's main activities are hip-hop street dance, writing a hip-hop-inspired blog and a generally hip-hop-related lifestyle. Hanne's interests include horseriding, photography and blogging. Their blogs mirror their creativity and interests. Ugur's main interest is digital technologies. He attends a Turkish weekend and evening school organised by a Turkish organisation which focuses on immigrants' living conditions by way of strengthened education through a focus on digital technologies, maths, Norwegian and Turkish culture and languages and performances.

Portrait 1: Ugur

Background

In the first interview, Ugur explains that he lives a happy and active life with lots of friends in the Grorud Valley, and his family has a large ethnic and extended family network. Ugur is interested in computers, computer games, social media, maths, sports and his youth club. He has a computer in his bedroom and is proud of a toy helicopter he has constructed. He spends a lot of time using his iPhone (interview, spring 2011). In his final interview, he mentioned looking forward to building a new robot at the out-of-school Turkish Centre (interview, spring 2012).

Digital media and interests as funds of knowledge

At the out-of-school Turkish Centre, Ugur involves himself with mathematics and Lego Robotics. He ranks the teachers as very good; most of them are Turkish bilingual university students. He reports:

> You have clever people, engineering students, technology students (he looks proud). You are one of two or three youngsters getting help from one student in a very small classroom. Very good! Very good! I joined the maths class to become as good as my older brother in maths. (interview, spring 2011)

During participant observation in the maths class at his upper secondary school, Ugur plays iPhone games during class activities while commenting:

> The maths level here is basic. Really basic! I cannot be bothered working with maths in school any more. That's why I play computer games, you know? [whispers while looking at the teacher] I practise at the Centre ... don't tell my parents [laughs]. (interview, autumn 2011)

Ugur joined the technology courses at the Turkish Centre because he remembered having fun building a steam engine with some Lego Technics he received as a present from his mother in his childhood. An interest started to develop; he spent time exploring the internet for manuals, built model helicopters and became generally interested in the subject. He discovered that he could solve and manage technological challenges in a research-based manner. Simultaneously, he developed a competence for using computers, both software and hardware (interview, spring 2011). In primary school, friends and teachers started to ask him for help, and he has developed a role as a computer expert in his social network (school observation, spring 2011).

Without having deep understanding of electricity or electronics per se, his experience with Lego Robotics, for example, empowers him to experiment and solve problems. Figure 5.1 shows how he has connected television antennae, wireless networks and such at home by logically understanding how they should be assembled based on colour codes, socket

Figure 5.1 An example of Ugur's handiwork in connecting networks, television antennae and so forth based on knowledge gained from playing with Lego Robotics and such tools

Source: Image taken by Ugur, 2011.

design and use of manuals and illustrations found on the internet.

Ugur's interests and competences with computers have also transferred to other activities. He made, on his own initiative, a video for the graduation ceremony at his secondary school based on video clips and pictures taken with his iPhone during the school year. The movie was a huge success (observation at the graduation ceremony, summer 2011).

In his edited movie, Ugur seems inspired by the road movie genre with his use of the escape metaphor related to leaving school and the silent movie genre with a nod toward Chaplinesque comedy.

Figure 5.2 **Picture sequence from Ugur's movie**

Source: Images taken by Ugur

On the subject of his success, Ugur explains:

> It is nice when people applaud your work. You get attention, you get respect ... I became very happy, you know? People begin to say hello to you. You get new friends. This does something with your self-esteem, you know? You can do something, you know? It is nice when people applaud your work. I know that I can if I want! (interview, summer 2011).

Ugur's interest in digital technologies has paid off in a formal context. Recently, Ugur and some classmates won second prize at a prestigious entrepreneurial competition in Oslo for upper secondary students. The group combined electronics and digital devices to create a standby-mode, energy-saving device using knowledge from the work Ugur had done at the out-of-school Turkish Centre and his own experiments at home (interview, summer 2012).

Learning identity

Ugur's development of skills and competencies in digital technologies has given him a methodological and research-based attitude; he can be characterised as solution-oriented and reason-based in his approach to tasks both inside and outside of school. Ugur takes advantage of this methodological way of working in maths, science and project work at school. It may be that these subjects per se allow this kind of work methodology to a larger degree than in other subjects. His self-initiated, out-of-school practices have had a major impact on his performance at school.

During the interviews, it becomes clear that the experiences and technological skills that Ugur has developed provide self-esteem and shape his learning identity. His grades are average in some subjects, but his interests in digital technology and maths give him status and confidence. His belief in himself characterises his learning identity.

Ugur's ambition is to become an accountant, and his dream is to return to Turkey to establish a profitable business; he describes Turkey as less bureaucratic than Norway (interview, spring 2011). This ambition can be interpreted in relation to both family-based funds of knowledge showing that one can leave one country (Turkey) to make a living in another and that by working methodically one can realise one's goals. His networked life has given him self-confidence in the sense that anything is possible.

Portrait 2: Hanne

Background

Hanne and her sister moved in with her grandmother and step-grandfather in the Grorud Valley when her mother died some years ago. After her mother died, Hanne began horseriding and photography. She found inspiration from her mother who had an interest in horses and worked as a self-taught photographer in a photography shop. Her visits to the photography shop and close examination of her mother's pictures became especially important for Hanne. Hanne draws on her interests when photographing horses in their natural environment. When horseriding or taking photos she feels free and at one with nature. She describes herself as 'normal', but exclaims laughingly that 'girls who like horses aren't that popular' (interview, spring 2011).

Digital media and interests as funds of knowledge

Hanne inherited an advanced SLR camera from her mother and bought a computer and an iPhone. She has worked hard

Figure 5.3 **Two of Hanne's photos of horses in their natural environment**

Source: Image taken by Hanne, 2011.

to improve her photography skills, which she uses to focus on horses. She explains how difficult it used to be to capture horses in their natural environment, 'the setting that makes you feel free' (interview, spring 2011).

Her interest in photography and horseriding has led to an interest in blogging about horses; blogging also works as a networking space. Throughout secondary school she blogged daily and commented on other blogs. Her blog contains images and video clips of horses, horseriding and her personal diary. Blogging became an informal way of practising and improving reading and writing. Her Norwegian grades in school improved significantly due to this because 'a good blog cannot contain misspelling and poor writing', she explains. She describes this work as very rewarding (interview, spring 2011). Through blogging she has developed a general interest and skills in computer technologies, electronic gadgets and internet use. Her self-initiated skills have resulted from her research-based learning using the internet itself as a manual (interview, autumn 2011).

Figure 5.4 illustrates Hanne's out-of-school interests in photography, horseriding, video editing software and blogging.

Figure 5.4 Two screen shots from Hanne's horse blog

In secondary school, Hanne sometimes struggled with academic motivation, but a strong desire to enter the media and communication vocational programme in upper secondary school inspired her school work. Her interests in digital media and related formal school subjects results from her interest in photography and blogging. Her dream in secondary school was to become a skilled photographer (interview, summer 2011).

As part of studying media and communication in upper secondary school, Hanne has acquired competences in using software, and graphic design is her favourite topic. Good graphic design demands a tidy and schematic approach; composition and use of colours trigger her creativity, and with her mastery of digital technologies has come confidence. These subjects have given her literacy practices that improve her work in school (interview, summer 2012).

Learning identity

Hanne develops informal and formal competences using computers, her iPhone and different types of software. Digital media function as a strong catalyst for developing her learning identity, thus enabling Hanne to use knowledge from informal learning situations to improve learning in school. When using her informal media competences in

Figure 5.5 Examples of Hanne's graphic designs from her media and communication classes

school, she believes that she can master subjects important for her future orientation.

Hanne's future orientation is currently leaning more toward graphic design than photography because she handles this topic well. She plans to live in the Grorud Valley working with graphic design (interview, spring 2012).

Portrait 3: Aisha

Background

Aisha's mother died some years ago, and Aisha and her sister have little contact with their father. They live with their uncle who has custody over them. Like many others in her family, Aisha's mother left Gambia to get a job in a Western country. She settled in the Grorud Valley where Aisha was born. Her mother valued traditional and modern fashion, proper conduct, international travel and relationships, making traditional meals, and strong family bonds. It was important for her mother to transfer these typical Gambian values to Aisha, as she explains herself, 'my mother transferred this knowledge to me before she died' (interview, spring 2011).

Dance is an important part of Aisha's life. She takes lessons and instructs and participates in hip-hop street-dance performances. Hip-hop dance, fashion and lifestyle are highly valued competences and community-based funds of knowledge among her friends. Aisha is Muslim, but she is allowed to stay out with friends and go to parties. Several of the other girls in her class in secondary school thought that she had a 'bad-girl' attitude and that 'she was allowed to do "too much"' (interview, spring 2011).

Digital media and interests as funds of knowledge

Aisha mentions fashion, travel, cultural diversity and taking care of her sister as her main interests inspired by her mother, and thus these represent facets of her fund of knowledge. She has strong family bonds and a network with her relatives in London and New York. She routinely uses social media to talk and chat with her relatives and spends her summer vacations with relatives abroad. This gives her insight into different cultures (interview, spring 2012).

She is much occupied with the African American hip-hop lifestyle popular in the Grorud Valley. Important hip-hop values that appeal to Aisha include showing social acceptance and being an inclusive person (interview, spring 2012).

Her interest in hip-hop introduced Aisha to blogging, with inspiration coming from French and US hip-hop blogs. One day, having nothing to do, she started to browse the internet for an introduction on how to make a proper hip-hop blog (predominantly picture-based with some video clips).

Figure 5.6 **Examples of traditional, normal (relaxed) and modern hip-hop clothes**

Source: Pictures taken by Aisha's friend in London (summer 2011)

On YouTube she found a step-by-step manual. Creating her own blog proved quite difficult, but successfully accomplishing the task gave her great satisfaction. Aisha says the blog portrays her. It contains links to all the stars, videos, music, fashion and religious items important to her – especially the hip-hop lifestyle and traditional Gambian culture. The blog has inspired her creativity and given her practical computer skills. She has worked in a research-based way to develop these skills (interview, spring 2012).

In the services and businesses programme at upper secondary school, Aisha pursues the interests and funds of knowledge transferred to her by her mother. She draws on her cultural experiences and orientation as the subjects often contain links to different cultures. She can use her digital skills in several of the subjects. Most of her skills have been developed outside of school and have become relevant in the subjects taught in school. For example, frequent visits to her family in England and the USA have improved her English grades significantly (interview, spring 2012).

Aisha sees a clear divide between secondary and upper secondary school in the way she can use her funds of knowledge and digital technologies. She described herself as insecure in secondary school, while she now masters the school subjects much better (interview, spring 2012).

Figure 5.7 illustrates Aisha's media competences in an upper secondary school advertisement project combining PowerPoint presentation tools, graphic design and cultural understanding. The slide is inspired by her blog and shows

Figure 5.7 **An illustration of noise, the slide vibrates slightly and shows Aisha's global cultural understanding and literacy practices**

Source: Aisha's advertisement project in upper secondary school (2012)

a picture from Times Square in New York, mixing different cultural expressions and resources. Using digital media makes it possible for her to combine different expressions and multimodal resources with which she feels comfortable. She experienced New York as a 'noisy' melting pot and used multimodal expressions to illustrate this (interview, spring 2012).

Learning identity

In upper secondary school, Aisha benefits from her interest in culture, travel and digital graphic creativity. Her profound interest in fashion, culture and travel has made her select an educational trajectory based on her funds of knowledge. Digital media have catalysed her interest in global cultures, which is important in her educational trajectory as she orients herself toward a service profession. Originally, she chose subjects in upper secondary school to work in the airport service sector. Now, however, her ambition is to become an air stewardess. To succeed she has concluded that she needs to learn an additional foreign language; she has chosen Spanish. Increased self-confidence has given her a learning identity and belief in success that makes her interested in school. Her future ambition is to leave the Grorud Valley and Norway to see the world and live abroad, possibly in New York (interview, spring 2012).

Interpretation of the portraits

The three portraits describe everyday 'learning lives' in the Grorud Valley and explore traversals of boundaries from informal to formal learning. Digital technologies serve as an integrated part of the youths' everyday lives inside and outside of school. The concept of network becomes important as an analytic lens for understanding the different interconnections and funds of knowledge the youths have in their learning lives and how these play together. The portraits show that digital media, as cultural tools, are key aspects for the links between informal and formal learning. We focus on how digital media support certain learning trajectories and learning identities for these young people rather than how digital media are network tools themselves.

The three participants shared computer-related competences that emerged from their everyday funds of knowledge-based personal interests such as building Lego Robotics, horseriding and hip-hop dance. These informal interests are based on funds of knowledge and relate directly to family backgrounds and cultural belonging. Such connections can be based on feelings (Barron, 2006), and these interests open up trajectories within formal learning settings.

The participants learned reasoning from the use of digital media when they had to put together different pieces of information (e.g. from the internet) to solve different tasks like creating new Lego Robotics systems or internet blogs from scratch. They learn the necessity of following procedures and rules carefully when making a professional video-clip layout or efficient programme elements in Lego Robotics or blogs. They understand that meticulously following certain procedures pays off in many school subjects, too.

The youths are also conscious about how they draw on interests and expertise from informal to formal contexts. The subjects they have chosen in upper secondary school and the way they solve tasks in school enable them to draw on these experiences when they create learning trajectories. Being conscious about their interests and abilities shapes their future orientation. For example, based on Hanne's interest in photography and horseriding, she has been able to publish some graphic work, which in turn has encouraged her to choose graphic design subjects in school.

Further, the three participants shared the experience that their informal digital practices positively transfer skills and knowledge that support their academic performance and development as learners. The portraits point to important individual outcomes expressed in the development of their learning identities. We can describe these as ontological aspects of learning as opposed to epistemological aspects (cf. Packer and Goicochea, 2000; Wortham, 2005). Learning is not only about what you learn but also about how you change as a person as you learn within different contexts.

We draw some important insights from this. For example, the students' interest in digital technologies and the ability to use them in school are based on their personal interests developed from funds of knowledge. They have acquired research-based and structured methods while performing leisure-time activities as self-initiated learning. The structured approaches learnt at home serve the students well in the school context. A 'networked life' builds self-confidence and a stronger belief in one's own future orientation as a learner. This reflects necessity and usefulness of understanding as to what kind of knowledge students possess and develop as out-of-school learners (Gonzales et al., 2005; Barron, 2006).

In our research, self-sustained learning can begin as interplay between digital media and funds of knowledge. Self-initiated learning, based on personal interests, offers a learning identity that enables youths to think positively about learning and themselves as learners in a formal setting. In the cases of our three students, digital media have catalysed this process. This was almost impossible in secondary school, but in upper secondary, youths can choose courses where they can draw on their informal competencies. It is therefore important for the school to understand and utilise students' informal competences in the education process (Krange and Øya, 2005).

Conclusion

Our ethnographic biographical study investigates how the learning identity of three 15-year-old students benefitted from the use of digital media and digital technologies. We show strong indications that during their first year in upper secondary school, these students' interest in school improved as a result of their choosing subjects that cohered with their out-of-school interests.

The holistic approach shown in this chapter offers a way of bridging the binary opposition between formal and informal learning and underpins the debate about how home, community, informal contexts and school may be reinscribed as changed and changing sites of learning (Sefton-Green, 2004). It contradicts the idea that only certain types of family-based cultural capital create school 'winners'. Our findings illustrate that schools must strategically emphasise and identify informal funds of knowledge related to the use of digital media in formal learning. Such preconditions can enhance learning if education builds on such funds.

From this we can conclude that networked lives for learning represent the new platform of which the interconnections between learning and use of digital media across formal, non-formal and informal contexts need to be explored and understood as we approach learning for the twenty-first century.

Note

1. The Local Literacies and Community Spaces project involves in-depth analysis of 60 informants on different levels, see: *http://www.uv.uio.no/pfi/english/research/projects/erstad-learning-lives/index.html* (accessed 3 September 2012).

Bibliography

Barron, B. (2004) 'Learning ecologies for technological fluency in a technology-rich community', *Journal of Educational Computing Research* 31: 1–37.

Barron, B. (2006) 'Interest and self-sustained learning as catalysts of development: a learning ecology perspective', *Human Development* 49: 193–224.

Bekerman, Z., Burbules, N.C. and Silberman-Keller, D. (2006) 'Introduction', in Z. Bekerman, N.C. Burbules and D. Silberman-Keller (eds), *Learning in Places. The Informal Education Reader*, New York: Peter Lang, pp. 1–8.

Bourdieu, P. (1986) 'The forms of capital', in J. G. Richardson (ed.), *Handbook of Theory and Research for the Sociology of Education*, New York: Greenwood Press, pp. 241–58.

Bråthen, M., Djuve, A.B., Dolvik, T., Hagen, K., Hernes, G. and Nielsen, R.A. (2007) *Levekår på vandring. Velstand og marginalisering i Oslo [Living Conditions in Transition. Welfare and Marginalization]*, Oslo: Fafo Report.

Bruner, J.S. (1994) 'The remembered self', in U. Neisser and R. Fivush (eds), *The Remembering Self: Construction and Accuracy in the Self-Narrative*, Cambridge: Cambridge University Press, pp. 41–54.

Buckingham, D. and Willett, R. (eds) (2006) *Digital Generations. Children, Young People, and New Media*, Mahwah, NJ: Lawrence Erlbaum.

Castells, M., Linchuan Qui, J., Fernandez-Ardevol, M. and Sey A. (2006) *Mobile Communication and Society. A Global Perspective*, Boston, MA: MIT Press.

Dreier, O. (2003) 'Learning in personal trajectories of participation', in N. Stephenson, H.L. Radtke, R.J. Jorna and H.J. Stam (eds), *Theoretical Psychology. Critical Contributions*, Concord, ON, Captus University Publications, pp. 20–9.

Edwards, R. (2009) 'Introduction: life as a learning context?' In R. Edwards, G. Biesta and M. Thorpe (eds), *Rethinking Contexts for Learning and Teaching. Communities, Activities and Networks*, London: Routledge, pp. 1–13.

Edwards, R., Biesta, G. and Thorpe, M. (eds) (2009) *Rethinking Contexts for Learning and Teaching. Communities, Activities and Networks*, London: Routledge.

Erstad, O. (2010) *Digital kompetanse i skolen, 2. utgave* [*Digital Competence in the School*], 2nd edn, Oslo: University Press.

Erstad, O. and Gilje, Ø. (2008) 'Regaining impact – media education and media literacy in a Norwegian context', *Nordicom Review* 29(2): 219–30.

Erstad, O. and Quale, A. (2009) 'National policies and practices on ICT in education: Norway', in T. Plomp, R. Anderson, N. Law and A. Quale (eds), *Cross-national Information and Communication Technology Policy and Practices in Education*, Greenwich, CT: Information Age Publishing, pp. 341–52.

Erstad, O., Gilje, Ø. and de Lange, T. (2007) 'Re-mixing multimodal resources: Multiliteracies and digital production in Norwegian media education', *Journal of Learning, Media and Technology* 32(2): 183–99.

Gee, J. P. (2003) *What Video Games Have to Teach Us about Learning and Literacy*, New York: Palgrave Macmillan.

Gonzalez, N., Moll, L.C. and Amanti, C. (eds) (2005) *Funds of Knowledge. Theorizing Practices in Households, Communities, and Classrooms*, Mahwah, NJ: Lawrence Erlbaum.

Holland, D., Lachicotte, W., Jr., Skinner, D., and Cain, C. (2003) *Identity and Agency in Cultural Worlds*, Cambridge, MA: Harvard University Press.

Hull, G.A. and Greeno, J.G. (2006) 'Identity and agency in nonschool and school worlds', in Z. Bekerman, N.C. Burbles and D. Silberman-Keller (eds), *Learning in Places: The Informal Educational Reader*, New York: Peter Lang, pp. 77–97.

Hull, G. and Schultz, K. (2001) 'Literacy and learning out of school', *Review of Educational Research* 71: 575–611.

Ito, M., Baumer, S. Bittanti, M., Boyd, D., Cody, R., Herr-Stephenson, B., Horst, H.A., Lange, P.G., Mahendran, D., Martinez, K.Z., Pascoe, C.J., Perkel, D., Robinson, L., Sims, C. and Tripp, L. (2010) *Hanging out, Messing around, and Geeking out*, Cambridge, MA: MIT Press.

Kozma, R.B. (ed.) (2003) *Technology, Innovation and Educational Change. A global Perspective*, Eugene, OR: International Society for Technology in Education (ISTE).

Krange, O. and Øya, T. (2005) *Den nye moderniteten. Ungdom, individualisering, identitet, mening [The New Modernity. Youth, Individualization, Identity, Meaning]*, Oslo: Cappelen.

Latour, B. (2005) *Reassembling the Social – An Introduction to Actor-Network-Theory*, Oxford: Oxford University Press.

Lave, J. and Wenger, E. (1991) *Situated Learning: Legitimate Peripheral Participation*, Cambridge: Cambridge University Press.

Law, N., Pelgrum, W. J. and Plomp, T. (2008) *Pedagogy and ICT use in Schools around the World. Findings from the IEA SITES 2006 Study*, Hong Kong: Springer Comparative Education Research Centre, University of Hong Kong.

Leander, K.M. and McKim, K. K. (2003) 'Tracing the everyday "sitings" of adolescents on the internet: A strategic adaptation of ethnography across online and offline spaces', *Education, Communication & Information* 3(2): 211–40.

Leander, K.M., Phillips, N.C. and Taylor, K.H. (2010) 'The changing social spaces of learning: Mapping new mobilities', *Review of Research in Education* 34: 329–94.

Lemke, J. (2007) 'Identity, development, and desire: Critical discourses and contested identities', in C.R. Caldas-Coulthard and R. Iedema (eds), *Identity Trouble: Critical Discourse and Contested Identities*, New York: Palgrave Macmillan, pp. 17–42.

Livingstone, S. (2009) *Young People and New Media*, London: Sage.

McLeod, J. and Yates, L. (2006) *Making Modern Lives. Subjectivity, Schooling and Social Change*, New York: State University of New York Press.

Mishler, E. (1996) 'Missing persons: Recovering developmental stories/histories', in A. Colby and R. A. Shweder (eds), *Ethnography and Human Development: Context and Meaning in Social Inquiry*, Chicago, IL: University of Chicago Press, pp. 73–100.

Moje, E.B., Ciechanowski, K.M., Kramer, K., Ellis, L., Carrilo, R. and Collazo, T. (2004) 'Working toward third space in content area literacy: an examination of everyday funds of knowledge and discourse', *Reading Research Quarterly* 39(1): 38–70.

Moll, L.C., Velez-Ibanez, C., Greenberg, J., Rivera, C., Andrade R. and Dworin, J. (1990) *Community Knowledge and Classroom Practice: Combining Resources for Literacy Instruction*, Technical report (OBEMLA Contract No. 300-87-0131), Tucson, AZ: University of Arizona College of Education and Bureau of Applied Research in Anthropology.

Nielsen, H.B. (2009) *Skoletid* [*School Time*], Oslo: University Press.

Nordicom (2009) *Young People in the European Digital Media Landscape. A Statistical Overview with an Introduction by Sonia Livingstone and Leslie Haddon*, Gothenburg: Nordicom, University of Gothenburg.

Norwegian Media Authority (2008) 'Safe use – a survey of 8 to 18 year olds' use of digital media 2008', available at: *http://www.medietilsynet.no/no/Trygg-bruk/Ressurser/* (accessed 1 April 2011).

Nunes, M. (2006) *Cyberspaces of Everyday Life*, Minneapolis, MN: University of Minnesota Press.

Packer, M. and Goicoechea, J. (2000) 'Sociocultural and constructivist theories of learning: ontology, not just epistemology', *Educational Psychologist* 35: 227–41.

Sawyer, R.K. (ed.) (2006) *The Cambridge Handbook of the Learning Sciences*, Cambridge, MA: Cambridge University Press.

Sefton-Green, J. (2004) *Report 7: Literature Review in Informal Learning with Technology Outside School*, Bristol: Futurelab.

Selwyn, N. (2011) *Schools and Schooling in the Digital Age. A Critical Analysis*, London: Routledge.

Thomson, P. (2011) *Unfolding Lives. Youth, Gender and Change*, Bristol: The Policy Press.

Vygotsky, L.S. (1978) *Mind and Society: The Development of Higher Psychological Processes*, Cambridge, MA: Harvard University Press.

Wenger, E. (1998) *Communities of Practice: Learning, Meaning and Identity*, New York: Cambridge University Press.

Wortham, S. (2005) Learning Identity: The Joint Emergence of Social Identification and Academic Learning, Boston, MA: Cambridge University Press.

6

Network and mobile technologies in education: a call for e-teachers

Guglielmo Trentin

Abstract: There are still barriers hindering the widespread use of network and mobile technologies (NMTs) at school. The main issue is the diffusion of the new pedagogical approaches required to best exploit that technology for the improvement of learning processes. Take for example so-called 'e-pedagogy', which relies greatly on learning-by-doing, connectivity and the strong integration of formal education with the informal learning processes which are typical of peer-learning, experiential learning, etc. At least two conditions favour wide propagation of e-pedagogy: (a) an organisational development of the school institution to foster didactic innovation in which NMT use is the normal condition, just as use of the blackboard, paper and pen were for a long time the normal condition; and (b) a process of continuous professional development of teachers which is not only based on formal training so much as on informal learning processes specifically centred on NMT use. To act on both these conditions we must be aware of the rapid changes in the context and consequently of how the figure of the 'e-teacher' should be modelled to produce someone who is able to use NMTs both for their students' learning and for their own continuous professional development.

Keywords: e-teacher, 2.0 teaching, e-pedagogy, continuous professional development, pedagogical innovation

Introduction

Even if the diffusion of network and mobile technologies (NMTs) is bringing about marked changes in the way individuals interrelate, the gap between how students and schools use them is ever-widening. Furthermore, as the teaching/learning process is based above all on communication and social interaction, people involved with education cannot ignore these changes; indeed, they must adjust to them (Kearsley and Blomeyer, 2004).

It is thus evident that the school system must adopt a strategic role both in guiding the new generations to an ethically correct and discerning use of communication technology, and in proposing it as a tool for enhancing individual learning processes, extending them outside the physical and institutional confines of school (Faberman, 2005; Barnes et al., 2007; Rocha, 2007; Silva, 2007).

As teachers are the driving force behind teaching and learning activities developed at school, it is inevitably up to them to initiate a process of change in pedagogy. This change must take into account the multiplicity of information and interaction channels that students have daily at their disposal. This is an extraordinarily important task from the educational point of view, one which requires a change both in the teacher's role and in didactic organisation. Such a change has been defined as 'epoch-making' (Fullan, 1997; Goodson, 2003), and it requires teachers to learn to teach differently from how they themselves were educated, as regards both the specific subject area and the teacher's role (Hargreaves et al., 2001; Gaston, 2006; Thompson, 2007).

It is for this reason that, if we wish to promulgate the knowledge, skills and culture related to the educational use of NMTs, we must use teacher training tools and approaches based on the same resources and methods by which they can then be proposed to students (Trentin and Repetto, 2012). This means no longer (or at least not only) formal training (i.e. participation in classroom or distance-learning courses), but interventions focused above all on informal learning processes (Cross, 2005; Cook, 2012), which exploit the potential of NMTs for accessing and sharing information, knowledge and good practices, by means of direct consultation of the online sources and social interaction in networked communities of practices (Wenger, 1998; Trentin, 2005).

It is certainly not an easy process, but it can succeed, especially if teachers are willing to invest time in a type of professional development aimed at tuning them into the communication channels of their direct users (the students) in order to exploit them for teaching and study support processes. This cannot of course happen if they do not familiarise themselves with NMTs. And the first step in this direction is generally to try and understand not so much how NMTs can be applied didactically (this will be acquired at a subsequent stage) but rather how to use them for one's own professional growth (access to knowledge, participation in online communities or training activities in which the same technological tools are used, etc.).

This chapter will focus on some of these aspects, contextualising the problems and trying to understand how the teacher's role changes when new methods of didactic communication and knowledge flow (Trentin, 2011) are introduced with NMTs. In other words, what conditions are required to make feasible a transition from traditional teaching to '2.0 teaching'? Furthermore, how should one model the figure of the 'e-teacher' to produce someone who is able to use NMTs for both students' learning and their own continuous professional development?

The key issues

The educational psychologist Steen Larsen (1986) argued that 'information can be transmitted but knowledge must be induced'. According to Larsen, three key stages bring about the flow of knowledge from the source (the teacher) to the receiver (the learner):

- *transformation of personal knowledge into public information*: teachers transform and organise their knowledge in order to transmit it to their students;
- *information transfer*: teachers transmit their knowledge, transformed into public information;
- *transformation by the students of the public information received into personal knowledge*: pupils transform the information provided by the teacher into personal knowledge.

Larsen uses a very effective metaphor for describing this inductive process: the electrical transformer (Figure 6.1). In

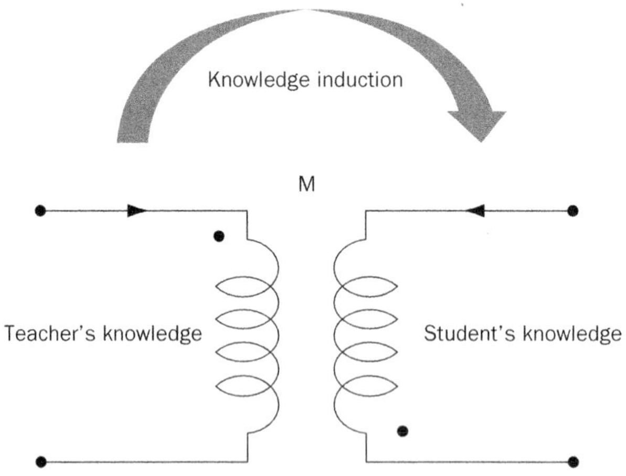

Figure 6.1 The electrical transformer metaphor

this, although there is no physical contact between the two circuits (the primary and the secondary circuits), a current produced by magnetic induction nonetheless passes between the two.

In this metaphor, the primary circuit of course corresponds to the teacher's repertoire of knowledge, and the secondary circuit to the learner's previous knowledge, which we are trying to develop further through induction.

The idea of induction also perfectly reflects our awareness that the mechanisms for the acquisition of new knowledge must not resemble the decanting of a liquid from one container (the teacher's head) to another (the student's head), but rather a process of absorption and integration of the information received by the student into his/her own pre-existing cognitive structures, which are the result of personal experience, earlier knowledge, etc.

Thus, one of the teacher's key roles is to create the conditions for stimulating and favouring this process of absorption and systematisation, by proposing both individual and collaborative learning activities, problem-solving and artefact development, etc. (Trentin, 2010).

School as one of the sources of knowledge, but not the only one

Until some time ago (certainly when the present teachers were students) the sources of information which could influence the development of learner's new knowledge, were limited to the teacher and subject-specific textbooks, in addition to media information channels such as radio and television; in any case all classifiable as so-called 'guaranteed' sources for their correctness and reliability.

The arrival of Web 2.0 (O'Reilly, 2006; Anderson, 2007) revolutionised this scheme, and for the student today there are many more channels for acquiring direct or indirect information in the various subject areas (e.g. Wikipedia). However, without adult control and support, transformation of this information into personal knowledge almost always involves a 'do-it-yourself' process wholly lacking in method.

The worrying aspect is that most teachers are unable to support this process (i.e. the favouring of what Larsen indicates as 'knowledge induction'), because they are still culturally anchored or induced by the school context to the use of conventional teaching/learning schemes centred on the teacher and the textbook (Coppola et al., 2002; Cochran-Smith, 2005).

There should be a parallel rethinking of the programming/organisation of teaching activities so that the use of technology is not so much an occasional event as normal practice. Evidently, individual teachers can do very little in this sense, and the margins of manoeuvre for each single institute are also very limited. As can easily be imagined, this fight must be fought at other levels, such as the ministries and policy-making bodies. It should not be forgotten, however, that if we wish to win the battle, we cannot ignore some fundamental tenets of organisation: 'if a foreign body is getting implemented in a system, either it adapts and will not be regarded as alien or it will continuously be identified as a foreign body and be eventually rejected from the system' (Euler and Wilbers, 2002; Seufert and Euler, 2003). In the present case, the 'foreign body' is clearly the NMT.

This is a key aspect which actually involves not only institutions but also each individual citizen, especially if we also consider the need for lifelong learning after complete formal education.

It is thus necessary to intervene on teachers' professional development, accustoming them to the informality of the learning processes offered by 2.0 resources.[1] This is a key aspect which indeed concerns not only the teaching process but also the continuous training of the individual teacher (Teras et al., 2010).

Regarding professional development, we should remember that growth both in the specific subject area and in teaching methodology will increasingly involve individual autonomy and the teacher's ability to master the communication technologies in daily use for his/her own continuous education (Hammond and Collins, 1991). This applies to the teacher today and will apply even more to the student in the future.

NMTs as a bridge between formal and informal learning

As discussed, the school has a pressing need to metabolise approaches to the educational use of NMTs: first, for the teachers, both as first-person users and as proposers of these approaches to their students; second, for the students themselves, for whom school may be the first context where they learn how to exploit methodically the full potential of NMT resources in the study of the various school subjects.

This is why teaching quality will be increasingly evaluated in terms of the education of the learner both in the individual use of network technology/services and in the capacity to become autonomous in providing for one's own continuous education in the specific contents domain, once the 'formal' learning process has been completed (Trentin, 2008).

In this sense, school can play a strategic role, not only educating but also 'meta-educating', i.e. equipping students to face the lifelong process of professional development (Hargreaves et al., 2001) – in other words, using the 'formal' educational process as a bridge, creating the necessary skills for subsequent autonomous management of 'informal' learning processes based on searching for and filtering information, accessing explicit (codified) knowledge sources, and navigating the ocean of social contacts trapped within the myriad of online social networks (for example, the networked communities of professionals).

Although these considerations are nowadays widely accepted, some basic critical questions remain, notably most teachers' poor knowledge of 2.0 resources and consequently their poor educational use of them.

The ever-widening gap

When discussing the integration of communication technology into teaching practices one cannot ignore the increasingly large gap between students' personal/daily use of NMTs and the way that teachers propose them for educational activities. On the one hand, there are the students (digital natives) (Prensky, 2001, 2004) with their innate capacity for mastering technologies, using them for both social and learning purposes, albeit with little or no method; on the other hand, there are the teachers (digital migrants) who, while making admirable attempts to innovate their teaching, tend towards conventional methods and practices of study, anchored to old teaching schemes, when proposing an educational use for NMTs. More appropriate would be to conceive and introduce new methodologies inspired by e-pedagogy (Elliot, 2008), which are able to exploit fully the potential of NMTs both for social interaction and for accessing information.

The widening of the gap is also exacerbated by the way institutions conceive of the physical spaces where the teaching/learning processes for integrating technology in education are supposed to take place.

The physical spaces of technology-supported education

If we look back in the recent history of educational technology, we can identify at least three stages in the long, slow process of introducing information and communication technologies (ICTs) into school teaching (Trentin and Repetto, 2012):

1. *The ICTs are installed and used inside a dedicated classroom.* The 'computer room' room has been organised specifically so that students can make periodical visits, sometimes to learn how to use the computers, and sometimes to use them in the study of other school subjects.
2. *With ICTs, particularly communication technologies, the classroom walls (of the computer room) are knocked down.* The computer is no longer seen as a tool to be programmed or for running educational software, but also as a powerful means both for accessing information, digital repertoires, and for coming into contact with other realities such as distant classes, other students, etc. For most students and teachers, however, the school's online connection remains their only means to access the internet.
3. *The classroom is extended into virtual space, fostering so-called extended/expanded learning* (Faberman and Kaplan, 2005; Rocha, 2007; Silva, 2007). With the diffusion of internet, both at home and in mobile forms (Wi-Fi technology, netbooks, tablets, smartphones, etc.), the computer room becomes obsolete, as the learning activities supported by the network can be developed anywhere: at school (not necessarily in a computer room), at home or on a park bench.

Now, although Stage 3 has already been in force for some years, schools and teachers are still trying to come to terms with Stage 2 (at least in Italy). In the meantime, students are increasing their use of social networks to interact with classmates, mostly for:

- activating somewhat unorthodox mechanisms for sharing and passing on solutions to exercises and tests;
- accessing information sources for study often with unfettered cutting and pasting from unreliable sites.

In other words: without any method.

The paradox is that someone else (the teachers) might teach them the method if only they had the minimum technological skills required to create the right mixture between technology and learning approach, enriched by the huge amount of resources to be found online.

The need for osmosis of reciprocal skills

What is needed, therefore, is a kind of partial osmosis of reciprocal skills, so as to rebalance them for an effective, efficient use of online resources in the learning of school subjects (Zimmerman and Trekles Milligan, 2007). How can this be achieved? Here the teacher must inevitably play a fundamental role, not only in his/her guise of subject area expert, but also in that of researcher (teaching implies a process of constant research) and even more in that of educator. This is possible only if the teachers are willing to:

- enter the communicative dimension of the new generations, using the students' own virtual spaces (i.e. 'going to visit' the students where they normally interact among themselves);

- indicate study methods which exploit the above dimension;
- educate students to use the potential of NMTs which are at their daily disposal in a discerning, intelligent way;
- educate students to digital citizenship.

In all this, we cannot ignore the urgent need for a systematic (and continuously updated) initial training programme for educators. Educators must be made aware of the need for change. They must recognise that the necessary change must come from within, and that this is the only way to achieve a partial alignment between the ways of communicating at school and in everyday environments (Zimmerman and Trekles Milligan, 2007).

We say 'partial alignment' because the educational mission of school should be to understand how new generations interact through NMTs, how they access web resources and interrelate socially; but it should also be to intervene where necessary to correct young people's flawed attitudes to the potential of technology and their wrong methods of exploiting it. In other words, teachers should educate young people to be digital citizens. In order to do this, however, teachers must themselves be good digital citizens, and here a vicious circle is often created.

At this point the question is: why should teachers (digital migrants, assuming they are willing to migrate) force themselves to enter into the mentality of a digital native and change their way of teaching, when school organisation itself is so alien to the demands of a 2.0 teaching method? In other words, if the teacher is required to follow the curriculum of their respective ministry of education, why make extra work (which in any case is usually not even acknowledged)?

The answer, as we said before, can only be found within each individual. Each teacher must decide whether to act merely as a 'content mediator', passing on knowledge of a specific school subject to his/her students (no small feat in itself), or whether to help them to become citizens of the future who are able not only to read, write and do arithmetic, but also to master methods and strategies for the effective and efficient use of communication technologies (Trentin and Repetto, 2012).

To prepare for this kind of educational mission, teachers must embark on a path with some compulsory initial steps, including:

- entering into the logic of the new communicative dimensions and practices induced by NMTs;
- understanding them fully by putting them into practice;
- mastering the 2.0 resources for their own professional development so as to be able to educate students to use them efficiently and discerningly in formal and informal learning processes.

What we are looking at is a path which enhances teachers' awareness of the potential of NMTs, starting from their first-person use before they are seen as teaching tools. This translates to a path attracting teachers to NMTs, by presenting them as a means to:

- access information, document archives, educational materials, digital repositories;
- contact and collaborate with colleagues and experts;
- take an active part in communities of practices;
- participate in online or blended teacher training events ('formal' education);
- practise continuous self-direct learning ('informal' learning).

After this first fundamental step, one must consider the way in which NMTs can be integrated into individual teaching practices; for example, how to integrate the 2.0 resources used daily by students to improve their approach to both individual and networked collaborative learning (Trentin, 2010).

This is an very delicate step, requiring teachers' strong awareness of how their role must change in the '2.0 classroom', i.e. a classroom which widens its range of action, extending it into virtual space, when learning takes place either in a physical classroom or outside the school building (Rocha, 2007; Silva, 2007).

Without a clear understanding of the required change in the teacher's role to accommodate the potential of the new communication channels and students' new ways of interacting, it will be impossible either to create suitable teacher training schemes (which are better applied early on), or to change the school's organisation so as to favour true didactic innovation based on the considerations expounded above.

Thus, the second part of this chapter will try to outline the profile of what might be defined as the 'e-teacher'.

From teacher to e-teacher

In e-pedagogy, the figure of the teacher continues to be central, although the function of the teacher changes as compared with exclusively classroom teaching. When teachers deliver a lesson they become facilitators in the process of learning content in which they are expert, contributing to the preparation of the teaching materials and/or supervising their students' interactive activities, both online and face-to-face. How far their function is modified obviously depends on the type of approach adopted each time, for example, whether it is more centred on the teacher or on collaborative study.

The knowledge and skills for carrying out this function efficiently, however, are not yet widespread among teachers (Robinson and Latchem, 2003). One element of e-pedagogy sustainability is thus closely linked to teacher professional development (Hudson, 2012), which should be concentrated not so much on the instrumental use of NMTs (training in the technologies), as on the various didactic/educational approaches for applying them to education.

The teacher's changing role

A 2.0 class requires both teachers and learners to take on different roles and responsibilities from a traditional class (Coppola et al., 2002; Crichton and Kinsel, 2003). Indeed, it has been some time now since Elliot (2008) called for the development of an e-pedagogy based on reconceptualisation of traditional teaching practices.

Teachers who intend to adopt NMTs in their practice therefore need to understand fully the philosophy underlying the concept of the 2.0 class and the paradigm shift it involves. Traditional teaching/learning practices are focused on the teacher, whose objective is to transfer a given body of knowledge directly to learners. By contrast, e-teaching concentrates on the relationship among learners, and that between learners and the knowledge to be acquired. Students are helped to be more autonomous, proactive and responsible with respect to their own learning processes.

Table 6.1 summarises and compares the main characteristics of traditional teaching (which may also make use of technology) and those related to the e-teaching paradigm.

Table 6.1 **Comparison of traditional teaching and e-teaching**

	Traditional teaching	E-teaching
Teacher	Possesses and transmits knowledge	Guides study
Student	Passively receives contents	Interacts with contents and the tutor/teacher, learns autonomously and in groups
Class	Place where knowledge is transmitted	Place where knowledge is built and shared
Knowledge and experiences	Transmitted vertically from teacher to learners	Shared horizontally among members of the learning group, which includes the teacher
Learning processes	Strongly directed	Active and collaborative
Curricular contents	Predefined and standardised	Flexible and open-structured so as to permit multiple learning paths
NMTs	Mainly used to support classroom lessons	The means for constructing an integrated, learner-centred learning environment
Teacher's perceptions of educational technology	Teacher considers educational technology as a surrogate for their own role	Technology is considered a means for stimulating learners, for improving and amplifying the learning environment

To recap then, systematic uptake of e-pedagogy is conditional on teachers having access to suitable professional development so that they become capable of making autonomous and informed decisions about what e-teaching strategies will prove most effective for meeting the needs at hand.

Moreover, given the particular features of online environments, which are certainly different from the physical environments where face-to-face lessons normally take place, the teacher must be trained to choose the most adequate strategies of interaction/didactic communication to fit the medium in which they are to be used.

This is a particularly critical competency for teachers; acquiring it should help dispel any impression they may have that the educational use of NMTs means simply transferring the contents and teaching approaches that have proved effective in the classroom onto the net. In short, the special characteristics of a given medium mean that it is never neutral in terms of communication dynamics and strategies (Trentin, 2010). For example, video, audio and multimedia each have their own characteristic pace and timing. Likewise, text communication, which is typical of electronic messaging systems (e-mail, forums, social networks etc.) where expressivity (tone of voice, interlocutor's expression etc.) is filtered by the medium often needs to be substituted with artifices such as so-called emoticons.

Before outlining the professional development path to be taken by a teacher who intends to adopt NMTs, we first need to define the abilities, knowledge and competencies required for effective implementation of e-pedagogy.

Enabling conditions for 2.0 teaching

As discussed, there may be many advantages in NMT-based teaching. There is just one key to adopting it efficiently, however: it is essential to believe in it.

As well as a firm belief in the potential benefits to learning, other factors indicated by Kearsley and Blomeyer (2004) are:

- the desire to innovate one's teaching approaches;
- the convenience of teaching where and when one wishes;

- the chance to establish a more personal rapport with individual learners;
- the host of materials available on the web;
- the possibility of teaching in a more stress-free environment.

Obviously, other factors can be influential, especially those regarding the teacher's status, e.g. the chance to earn more and to capitalise on e-teaching for professional growth and career advancement.

Willingness to engage in e-teaching

Once the teacher has sufficient conviction about the validity of online approaches, the next matter is the willingness and chance to modify teaching methods. In other words, what boundary conditions are to be considered for supporting e-teaching? Fuller et al. (2000) list the following demands made of the e-teacher:

- be willing to sit at the computer for at least 2–3 hours a day;
- feel comfortable with one-to-one interaction and debate;
- be flexible in teaching and interested in innovation;
- be willing to use multiple online services in the teaching/learning processes.

While these may not appear to be particularly stringent conditions, many excellent classroom teachers are unable to meet them. The reasons for this include difficulty in getting used to regular online interaction with learners, in adjusting to the demands of computer-mediated communication, and in feeling comfortable with technology.

Preconditions for e-teaching

From the above, we can derive some preconditions for becoming an effective e-teacher. In this regard, Kearsley and Blomeyer (2004) point to the need to:

- have ready access to the network;
- feel comfortable with tools and systems for teaching and learning online, such as social networks, wikis, e-learning platforms, authoring systems, etc;
- have prior firsthand experience of learning online – this is essential for understanding the potential advantages and pitfalls of NMT-based learning from the learner's viewpoint.

Online training is the best means of ensuring that the prospective e-teacher satisfies the last two preconditions. In this way, trainees will have gained learning experience using the same tools and methods that they will later adopt themselves in their own e-teaching activities (Trentin, 2010).

Cornerstones of e-pedagogy

One of the main conditions for effective e-teaching is full awareness of the cornerstones of e-pedagogy. These include:

- learner-centred activities;
- active, problem-based learning;
- problem-based learning related to daily experience;
- activities of information problem-solving, i.e. based on effective and efficient strategies for searching for information online;
- collaborative learning;

- teacher-role flexibility (facilitating individual learners and moderating learning groups);
- peer-based evaluation.

For many teachers, this is unfamiliar ground, so these approaches will need to be learned from scratch and integrated into their practice. Others may already have integrated them into their repertoire of classroom practice, but will nevertheless find it a challenge to implement them online.

Of the above strategies, the one that probably most characterises e-teaching is facilitation of individual and group learning (Collison et al., 2000; Salmon, 2002, 2004).

E-teacher education and professional development

In most European countries, initial teacher education is conducted in a traditional manner, with limited focus on the educational use of NMTs (Attwell, 2005) and teachers' capacity to innovate their teaching at the pedagogical and methodological levels.

The result is that after an initial phase of awareness-raising/training, the actual innovation process is left entirely up to individual teachers, who all too often find themselves on their own when it comes to introducing new solutions into teaching processes. To reduce this sense of isolation, conferences and seminars are sometimes held for disseminating news and good practices about educational NMT use. While useful, these represent only a partial solution to the problem. What is really needed is a new professional profile for teachers, and thus a reference framework for their continuing professional development (Attwell, 2005).

In other words, sustainable adoption of NMTs in support of teaching and learning processes should not be seen as an occasional, sporadic event but rather as part of a general change in teachers' status and in their continuing professional development.

What competencies does an e-teacher need?

As well as acknowledging the importance of the aforementioned points, we also need to recognise another key factor: definition of e-teachers' competencies as a guide to their professional development. There are still no generally-recognised standards for defining the competencies of the online teacher. However, some international bodies have made proposals in this direction. One of these is the International Society for Technology in Education (ITSE, 2010), which states that the e-teacher should:

- use social software competently;
- understand the characteristics of the 'e-learner';
- understand approaches and strategies for effective e-teaching;
- be fully aware of the pros and cons of computer-mediated communication;
- understand the dynamics of online group interaction and how to manage them;
- be capable of evaluating online activities (e-tivities);
- understand the legal (copyright) and ethical implications of online education;
- be aware of accessibility issues affecting disabled participants in online learning;
- understand strategies for effective integration of online activities into classroom learning.

This list is not exhaustive but it does cover the most important issues.

Besides defining the general guidelines for what the e-teacher's skills should be, we should consider another key aspect: their continuous professional development. Although this concern is common to all professional categories, it is undoubtedly particularly important for those professions based on intense knowledge flows (Trentin, 2011), experimentation and intensive use of technologies.

Need for continuous professional development

Teachers can be seen as artisans who work mostly alone, with a variety of (human) materials, organised in a highly personalised manner. They develop a repertoire of abilities and strategies to form an integrated set of different mental schemes which gains complexity over time. This repertoire is largely constructed through trial and error; indeed, teaching in the holistic sense also means experimentation. Thus, teachers are involved first-hand in constructing new knowledge and developing good professional practices.

In the last few years, their professional growth has become increasingly linked to school development processes. This is a strong stimulus for:

- formally defining the work that a teacher must carry out with his/her students;
- potentiating research, reflection and experimentation in work based on problem-solving;
- attentive examination of important research and experience regarding the best teaching practices;

- analysis of the effects of new communication technologies on students' learning styles;
- collaboration among teachers, preferably based on the activation of fully-fledged inter/intra-institutional communities of practice which exploit NMTs to amplify communication among their members and give it continuity.

In this way, solid foundations can be laid for the development of new approaches to managing and profiting from the immense heritage of tacit knowledge within the school, connecting research to professional development through the identification of good practices and the activation of inter/intra teacher and school networks.

The teacher's professional growth thus comes to depend more and more on their belonging to communities of practice whose main strategy is collaborative learning. This ongoing process of professional development requires reflectivity, dynamism and foresight on the teacher's part. Such qualities can only be fostered within a true culture of professional development, a culture that attributes importance to practice, experience and research, ideally conducted with peers and other professionals operating in the same context.

In this sense, teaching is a typical example of a profession with interdependence between practice and research. In other words, teachers draw on research but at the same time they can also be the source of new knowledge generated by their individual professional practice (Engestrom, 1995).

We can say then that the capacity to plan and conduct individual or group research into professional practice is part and parcel of teachers' professional development. Having those abilities also allows the individual teacher to evaluate research conducted by others, and to decide whether to apply it.

In this way, professional know-how becomes the result of combined knowledge gained from the workplace and the experience accumulated therein. In other words, professional knowledge is developed through a process of situated learning and is further elaborated by way of personal critical reflection (Hammond and Collins, 1991).

Implications for continuous professional development

Professional development thus appears to be based on a complex series of factors. The most obvious of these is the individual's direct involvement in their own professional development. However, there are numerous others, including (Attwell, 2005):

- exploration of, reflection upon and improvement of professional practice;
- development of the skills, knowledge and understanding necessary to evaluate and review professional practice;
- understanding of the processes of change (as practice increasingly takes place in complex and dynamic contexts);
- the ability to create new knowledge;
- development of theoretical knowledge to underpin and complement reflection upon practice;
- study of the interplay between theory and practice;
- the ability to transfer skills, knowledge and understanding from one context to another;
- the generation of expertise through research;
- the ability to handle complexity and inter-connectedness of issues (including through the formulation of mental models, schemas or networks);
- development of contextualised understanding;
- translation of understanding into action, as appropriate;

- further development of communication skills;
- the attempt to create a wider community of practice that embraces research as a guide to both policy and action;
- the ability to design and carry out authoritative research into aspects of professional practice;
- the ability to analyse, interpret, evaluate and, if appropriate, apply the research findings of others.

These conditions should provide the teacher with abilities and knowledge not just for using NMTs as a support tool, but for integrating NMTs into new pedagogical approaches (Thorpe, 2012).

We believe that all this is crucial to 2.0 teaching sustainability, and training projects such as Aladin 2.0 described in Chapter 7 of this book have in fact moved in this very direction.

Some conclusive reflections on e-teacher status

To integrate the use of NMTs into their teaching practices, teachers must change their attitude to teaching, shifting from a vertical model of knowledge transmission to a more horizontal one based on collaborative processes as well as on individual study. The teacher's role continues to be a central one, even if it is now rather as facilitator of the process than as mere dispenser of knowledge.

With this point of view, the nature of the teaching/learning process inevitably undergoes radical changes, as does the student–teacher relationship. The teacher becomes a less authoritarian figure, closer to the role of class intellectual resource, as well as facilitator of student activities. This situation tends to attenuate the authority dynamics which have traditionally existed in the classroom, and constitutes an undoubted opportunity for enriching the teacher–student relationship. To obtain these results, however, it is necessary to act on two strongly interrelated levels: professional development of teachers and formal recognition of e-teacher status.

Effective take-up of NMT-based learning is heavily reliant on the acquisition and spread of key competencies among teachers. The crux lies in understanding whether a rewards system might provide them with sufficient incentive to invest in development of those competencies, which are essential for e-pedagogy sustainability.

Another consideration concerns e-teacher status or rather how e-pedagogy competencies are regarded and certified. People qualified to teach are issued with a formal certificate documenting their subject-area expertise and their possession of the required academic qualifications. Certificates documenting NMT-based teaching competencies are rarely awarded similar recognition. These skills represent added value gained from experience and ongoing professional development at both methodological and technological levels. The absence of a proper recognition system offers a carte blanche to anyone wishing to try their hand at e-teaching, whether or not they have undergone adequate training. But what is worse, it generates an impression of e-teaching as a mere adjunct to classroom teaching, a fashionable accessory if you will.

This all points to two major conditions for supporting the sustainability of e-pedagogy, and hence of e-teaching:

- the need for general cultural growth regarding the attitude towards e-pedagogy (from the operators to the policy makers) and particularly towards its key figures, first and foremost the teachers;

- renegotiation of teacher/tutor status in their new role as e-teacher.

As we have seen, the former condition demands training courses designed specifically for developing the competencies needed to introduce NMTs into teaching/learning processes. The latter condition, on the other hand, requires in-depth reflection about how teachers' status varies when they adopt NMT-based methods.

As we have said, certain methods imply interaction with students and are not limited to classroom time but extend outside it. Such interaction is an activity which must be adequately recognised, formally redistributing online and onsite teaching hours. At present, however, as there is no official regulation of e-pedagogy; its adoption simply amounts to extra, unacknowledged, work for the teacher.

It may thus be concluded that for a true changeover to 2.0 forms of teaching to take place, a combined action is required: on the one hand, an organisational development of schools so as to include and make habitual new NMT-centred teaching/learning practices; on the other hand, formal recognition of the emergence of new professional figures without whom the sustainability of these practices could not be guaranteed. In the words of Euler and Wilber (2002), if this does not occur, these practices 'will continuously be identified as a foreign body and be eventually rejected from the system'.

Note

1. By the term '2.0 resources', we mean here both Web 2.0 technology and also what we have access to through that technology (information, factual materials, social networks, experts, etc.).

Bibliography

Anderson, P. (2007) 'What is Web 2.0? Ideas, technologies and implications for education', *JISC Technology and Standards Watch*, February, available at: *http://www.jisc.ac.uk/media/documents/techwatch/tsw0701b.pdf* (accessed 1 April 2011).

Attwell, G. (2005) 'E-learning and sustainability', *EdTechPost: Technology for Learning, Thinking and Collaborating*, available at: http:*//www.ossite.org/Members/GrahamAttwell/sustainibility/attach/sustainibility4.doc* (accessed 13 November 2010).

Barnes, K., Marateo R. C. and Ferris S. P. (2007) 'Teaching and learning with the net generation', *Innovate* 3(4), available at: *http://www.innovateonline.info/pdf/vol3_issue4/Teaching_and_Learning_with_the_Net_Generation.pdf* (accessed 22 June 2011).

Cochran-Smith, M. (2005) 'The new teacher education: for better or for worse?', *Educational Researcher* 34(7): 3–17.

Collison G., Erlbaum B., Haavind S. and Tinker R. (eds) (2000) *Facilitating On-line Learning: Effective Strategies for Moderators*, Madison, WI: Atwood Publishing.

Cook, J. (2012) 'Using educational technology to mediate informal task-conscious learning: design innovations in two European projects', *Educational Technology* 52(2): 39–43.

Coppola, N., Hiltz, S. and Rotter, N. (2002) 'Becoming a virtual professor: pedagogical roles and asynchronous learning networks', *Journal of Management Information Systems* 18(4): 161–89.

Crichton, S. and Kinsel, E. (2003) 'Learning plans as support for the development of learner identity: a case study in rural western Canada', *Journal of Adult and Continuing Education* 8(2): 31–6.

Cross, J. (2005) *Informal Learning*, Hoboken, NJ: Wiley Publications.

Elliot, B. (2008) 'E-pedagogy: does e-learning require a new approach to teaching and learning?', available at: *http://d.scribd.com/docs/22rc8wz72z067xrb1fpk.pdf* (accessed 23 November 2009).

Engestrom, Y. (1995) *Training for Change*, London: ILO.

Euler, D. and Wilbers, K. (2002) 'Selbstlernen mit neuen Medien didaktisch gestalten', in D. Euler and C. Metzger (eds) *Hochschuldidaktische Schriften*, St. Gallen: Institut für Wirtschaftspädagogik, pp. 127–42.

Faberman, D. and Kaplan, C. (2005) *Time for a Change: The Promise of Extended-time Schools for Promoting Student Achievement*, Boston, MA: Massachusetts 2020.

Fullan, M (1997) 'Planning, doing and coping with change', in A. Harris, N. Bennett and M. Preedy (eds) *Organisational Effectiveness and Improvements in Education*, Buckingham: Open University Press, pp. 19–26.

Fuller, D., Norby, R.F., Pearce, K. and Strand, S. (2000) 'Internet teaching by style: Profiling the on-line professor', *Educational Technology and Society* 3(2): 65–72.

Gaston, J. (2006) 'Reaching and teaching the digital natives', *Library Hi Tech News* 23(3): 12–13.

Goodson, I.F. (2003) *Professional Knowledge, Professional Lives: Studies in Education and Change*, Maidenhead: Open University Press.

Hammond, M. and Collins, R. (1991) *Self-directed Learning: Critical Practice*, London: Kogan Page.

Hargreaves, A., Earl, L., Moore, S. and Manning, S. (2001) *Learning to Change: Teaching beyond Subjects and Standards*, San Francisco, CA: Jossey Bass.

Hudson, B. (2012) 'Aiming for e-learning sustainability: transforming conceptions of teachers' professional e-learning', *Educational Technology* 52(2): 30–4.

International Society for Technology in Education (ITSE) (2010), available at: *http://www.iste.org* (accessed 1 November 2012).

Kearsley, G. and Blomeyer, R. (2004) 'Preparing teachers to teach online', *Educational Technology* 44(1): 49–52.

Larsen, S. (1986) 'Information can be transmitted but knowledge must be induced', *PLET* 23(4): 331–6.

O'Reilly, T. (2006) 'Web 2.0 compact definition: trying again', available at: *http://radar.oreilly.com/archives/2006/12/web_20_compact.html* (accessed 8 April 2007).

Prensky, M. (2001) 'Digital natives, digital immigrants', *On the Horizon* 9(5): 1–6, available at: *http://www.marcprensky.com/writing/Prensky%20-%20Digital%20Natives,%20Digital%20Immigrants%20-%20Part1.pdf* (accessed 4 September 2006).

Prensky, M. (2004) 'The emerging online life of the digital native: what they do differently because of technology, and how they do it', available at: *http://www.marcprensky.com/writing/Prensky-The_Emerging_Online_Life_of_the_Digital_Native-03.pdf* (accessed 4 September 2006).

Robinson, B. and Latchem, C. (eds) (2003) *Teacher Education through Open and Distance Learning*, New York: Routledge Falmer.

Rocha, E. (2007) 'Choosing more time for students: the what, why, and how of expanded learning', Center for American Progress, available at: *http://www.americanprogress.org/issues/2007/08/pdf/expanded_learning.pdf* (accessed 16 September 2008).

Salmon, G. (2002) *E-tivities: The Key to Active Online Learning*, London: Kogan Page.

Salmon, G. (2004) *E-moderating: The Key to Teaching and Learning Online*, London: Kogan Page.

Seufert, S. and Euler, D. (2003) 'Sustainability of eLearning innovations: findings of expert interviews', available at: *http://www.scil.ch/publications/docs/2003-06-seufert-euler-sustainability-elearning.pdf* (accessed 14 March 2005).

Silva, E. (2007) *On the Clock: Rethinking the Way Schools Use Time*, Washington, DC: Education Sector, available at: *http://www.educationsector.org/usr_doc/OntheClock.pdf* (accessed 28 September 2009).

Teras, H., Myllylä, M. and Nevalainen, T. (2010) 'Teacher training in the knowledge society: Web 2.0 and new professionalism', in J. Sanchez and K. Zhang (eds) *Proceedings of World Conference on E-Learning in Corporate, Government, Healthcare, and Higher Education 2010*, Orlando, FL, 18–22 October, pp. 1243–8.

Thompson, J. (2007) 'Is Education 1.0 ready for Web 2.0 students?' *Innovate* 3(4), available at: *http://www.innovateonline.info/index.php?view=article&id=393* (accessed 4 December 2010).

Thorpe, M. (2012) 'Does pedagogy still matter?' *Educational Technology* 52(2): 10–14.

Trentin, G. (2005) 'From "formal" to "informal" e-learning through knowledge management and sharing', *Journal of e-Learning and Knowledge Society* 1(2): 209–17.

Trentin, G. (2008) 'E-learning and teaching quality', *International Journal of Instructional Media* 35(1): 27–37.

Trentin, G. (2010) *Network Collaborative Learning*, Oxford: Chandos Publishing.

Trentin G. (ed.) (2011) *Technology and Knowledge Flow: the Power of Networks*, Oxford: Chandos Publishing.

Trentin G. and Repetto M. (2012) 'Becoming a networked teacher: a case study of the Aladin Project in Italy', in Q. Lin (ed.) *Advancement in Online Education: Exploring the Best Practices*, Vol. 1, Hauppauge, NY: Nova Science Publishers, pp. 47–63.

Wenger, E. (1998) *Communities of Practice: Learning, Meaning and Identity*, Cambridge University Press.

Zimmerman, L. and Trekles Milligan, A. (2007) 'Perspectives on communicating with the net generation', *Innovate* 4(2), available at: *http://www.innovateonline.info/index.php?view=article&id=338* (accessed 4 December 2010).

7

Networked informal learning and continuing teacher education

Manuela Repetto

Abstract: As it is widely accepted that the educational system must change, we might assume that one of the key factors for its innovation is the promotion of reliable continuing teacher education (CTE) initiatives. If the main purpose of CTE is to facilitate transformative practice and professional autonomy, rather than to provide a means of transmission, teachers must change their attitude to teaching. New constructs of schooling and learning are emerging, also due to the rise and rapid diffusion of network mobile technologies (NMTs). This chapter describes Aladin, a project in which teachers perform a sequence of steps in order to become fully aware of the educational value of NMTs. It focuses on the informal approach based on action research adopted in this project and on the results achieved, offering some reflections as to which critical factors would hinder or help such an initiative in becoming a continuous education practice in teacher professional development.

Keywords: informal learning processes, continuing teacher education, personal learning environment, action research, self-directed learning

Introduction

The impact of network mobile technologies (NMTs) in people's lives provides compelling new models of continuing professional development to prepare teachers to meet the demands of school and society.

Research shows that compared with their pupils, teachers feel inadequate concerning NMT use (Condie et al., 2007; Morris, 2010; OFSTED, 2009). Today's students even differ from their peers of a few years ago; they think and work in different ways and should be engaged with new learning models, better adapted to their needs (McKinney and Wilson, 2008; Underwood and Banyard, 2011).

Teachers, like students, are not merely recipients of knowledge (Richter et al., 2011). They are becoming increasingly autonomous in the organisation of their own learning processes (Attwell, 2007), and in the definition of their educational goals and strategies. Thanks also to the availability of user-centred NMTs, teachers are becoming shapers of their own educational experience (Banyard and Underwood, 2008).

Although teachers and students have a common active role as shapers, a teacher engages with technology differently from a student. The former recognises the value of technology for education but is uncomfortable with it and frustrated by the current curricula (Banyard and Underwood, 2008); the latter is self-confident and spontaneous in the use of technology, but lacks method (Trentin, 2010; 2013).

The emergence of Web 2.0 allows teachers and learners to customise the web to their own needs (Underwood and Dillon, 2011). Ubiquitous computing is offering new opportunities for the educational use of ICTs (Attwell, 2007), becoming a catalyst for encouraging pupils and teachers to work together in innovative ways (Hennessy et al., 2005).

Learning management systems (LMSs) and virtual learning environments (VLEs), which we could consider as the main technologies used at school during the previous web generation, are currently perceived by students as means for reproducing traditional forms of instruction and for controlling their learning, embodying classrooms within software (Attwell, 2007). According to an extensive survey (Underwood and Banyard, 2008), VLEs are perceived to be insufficiently innovative even by teachers, who are concerned about the top-down model employed in the design of VLEs and the lack of authentic learning opportunities for pupils.

On the other hand, personal learning environments (PLEs), which can be considered the key technology of Web 2.0, have the potential to gather together all the different formal and informal learning contexts, integrating learning from school with learning in the wider context. In this sense, PLEs can bridge formal and informal learning (Attwell, 2007).

If the educational system does not become more up-to-date, it will become increasingly ineffective for many learners, as they will tend to build a learning environment for themselves (Underwood and Dillon, 2011), in which the formal system will occupy only a marginal position. In fact, the importance of informal learning is increasing: learning takes place in multiple contexts and is not provided by a single learning provider only (Attwell, 2007). The risk for younger people is that they will consider school as a turn-off, a learning provider which is irrelevant for their identities and their lives (Attwell, 2007), out-of-date and overtaken by alternative agencies attracting them away from traditional settings (Johnson et al., 2012).

Teachers should therefore become aware of the importance of adequate and continuous professional development, as well as of their crucial role in transforming their classroom into a '2.0 classroom' and in preparing students to become active citizens in tomorrow's society. It was precisely to raise this awareness and to support this transition within schools that the Aladin project was set up.

The Aladin project: general approach and activities

The ALADIN (ALto ADige Instructional Network) project was conceived to support teachers in the discovery of the new Web 2.0 universe by means of experimentation, encouraging a gradual, discerning and independent approach to this universe. The aim was pursued by promoting informal learning processes within the teachers' community of practice, with the support of the Bozen Gruppo di Supporto alla Didattica con le ICT (GDS), a group of local teachers expert in the educational use of technologies, and coordinated by two researchers from the Institute for Educational Technology of the Italian National Research Council (ITD-CNR). The intention was to (a) improve teachers' knowledge of the pedagogical and educational use of Web 2.0 resources and tools; and (b) increase their skills in developing and trialling projects based on those same resources (Trentin and Repetto, 2012).

The Aladin project was proposed to teachers of three levels of school (primary, and lower and upper secondary) in the South Tyrol. The only requirements for participation were habitual use (at least at a personal level) of ICTs, and availability of internet access.

Below, after a brief description of the general methodological approach adopted in the Aladin project, we detail its evolution over a whole school year. Considerable space will then be dedicated to considerations about the proposed activities and training programmes, and about the results of the initiative.

The methodological approach

Of the various universally-recognisable approaches to continuing teacher education, the 'training model' has been the most dominant form of continuing teacher education (CTE) for teachers. This model is based on skills development and implies a technocratic view of teaching. It is delivered by an expert who determines the agenda and puts the teacher in the passive role of recipient of knowledge (Kennedy, 2005). The major shortcoming of this model is its lack of connection to the classroom context (Kennedy, 2005).

Alternative approaches to the training model are the action research model (Burbank and Kauchack, 2003), encouraging teachers to act as researchers and thus to take an active role; and the communities of practice model (Wenger, 1998), in which teachers create meaning and a common understanding through a negotiation process. In both these alternative models, as Eraut (1994) argues, it is not only the type of professional knowledge being acquired which is important, but also the context through which this is acquired, which gives a sense to that knowledge and appropriately embeds it in the teacher's own practice.

Aladin endorsed a mix of the latter two approaches (action research and communities of practice), the key idea being to create *action-research* situations (Greenwood and Levi, 1998) leading to the definition and experimentation of pedagogical approaches based on 2.0 resources. The action-research processes concentrated on the preparation and sharing of projects which could be easily integrated into normal curricula. To this end, teachers were asked to think of subject-specific activities which could be developed over a limited period of time.

The approach to online resources and project development methodology was not, as already mentioned, based on a rigid, formal educational path, but on informal learning processes stimulated within the community of practice. As the final goal was to propagate a new way of teaching centred on the use of 2.0 resources and hence on the equal importance of all the actors in the educational process (both teachers and students), the teacher training programme was also based on these same principles. Thus we rejected the idea of a formal (classical) programme managed in top-down fashion and centred on set time schedules, contents and activities.

We adopted an informal approach which would allow a very specific, personalised action, sufficiently flexible to adapt to the different characteristics and needs of different users, and based both on their levels of expertise in the educational use of ICTs, and on the different interests they expressed regarding specific 2.0 technologies. This flexibility also had the aim of making it possible to adapt Aladin to the different, unforeseeable needs which might arise in the more advanced stages of the project, with the consequent possibility of diversifying the proposed educational paths.

This led to the division of the project into two distinct pathways: Aladin 2.0 ('basic' programme) and Aladin 2.1 ('advanced' programme) (Figure 7.1).

The aim of the first Aladin programme (Aladin 2.0) was initial contact with the educational use of Web 2.0, and it was divided into three key stages:

- *familiarisation* with the technological resources chosen for an initial approach to Web 2.0;
- *preparation* of simple ICT-supported educational projects;
- *trialling* in class of the educational activities planned in the previous stage.

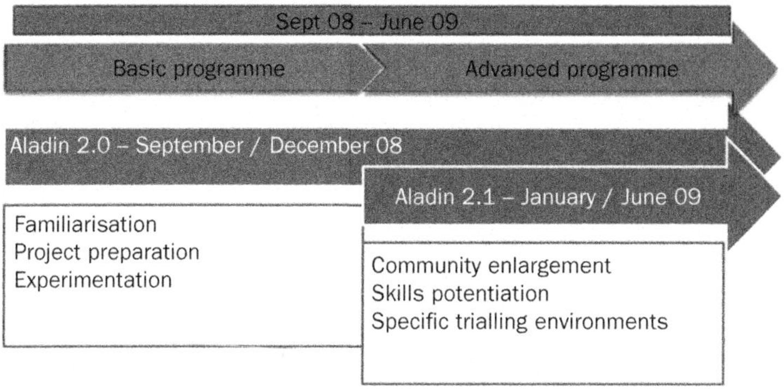

Figure 7.1 Time schedule for Aladin 2.0 and Aladin 2.1

While the aim of the first programme (Aladin 2.0) was to create the initial conditions for starting trialling in the classroom, the second (Aladin 2.1) aimed to:

- *extend the community* of teachers choosing to participate in Aladin, perfect the practices defined during Aladin 2.0, and repropose them to those who did not have the opportunity to try them out during the previous year;
- *potentiate the skills* of the teachers who had participated in Aladin 2.0, developing specific skills for assessing the impact of ICTs on their students' learning;
- *propose specific environments* for educational experimentation.

In this way, the more expert teachers had the chance, with the 'advanced' programme Aladin 2.1, to practise the basic steps learnt in Aladin 2.0 more thoroughly; the less expert teachers, or those interested in exploring a different Web 2.0 resource, were able to re-experience what had been done in Aladin 2.0 following other paths included in the 'basic' training programme for the educational use of Web 2.0.

The whole Aladin project was delivered in blended mode, alternating online activities and periodical face-to-face sessions consisting of moments in common and working sessions in subgroups. The distance activities, in particular, were organised within two specific virtual spaces set up on the Moodle platform, one for each programme of the Aladin project; these were modelled on the features of the two communities of teachers. The structuring of the Aladin virtual environment into two distinct sub-environments did not however bar the more expert teachers (of Aladin 2.1) from having access to the Aladin 2.0 environment. Indeed, this was explicitly suggested by the project coordinators to provide peer support to the less expert colleagues.

The actors in the process

The various professional figures involved in the project were:

- The ITD researchers, who planned and coordinated the Aladin programmes in collaboration with the Bozen School Board;
- the board support group (GDS), made up of eight teachers expert in ICT, each of whom was assigned to a group of schools (organised by geographical area and school level) and provided on-the-spot and online support connected to the proposed 2.0 resources;
- the teachers of the three school levels involved in Aladin, organised into two groups: the 'juniors' (50 people), who participated in the Aladin 2.0 programme; and the 'seniors' (25 people), almost all originating from Aladin 2.0, who were involved in the subsequent programme, Aladin 2.1.

As specified earlier, the participants in the Aladin project, besides meeting one another in person during the face-to-face events, also used the web to maintain remote contact, through the socialisation forums, the forums dedicated to the use of specific 2.0 web resources, and the discussion areas set up for project planning. These contacts offered further opportunities for small working groups to form spontaneously around common interests linked to a specific 2.0 resource, a particular subject area, or a particular school level.

The web was also seen to be particularly useful for overcoming not only geographical distances (the researchers are based in Genoa, about 600 km away from Bolzano) but also teachers' difficulties in moving around the Dolomite territory of South Tyrol, which is particularly mountainous and snowbound for a large part of the year.

The support environments for 'informal' training

The criteria guiding the setting up of the virtual environments for Aladin considered teachers' skills levels, their 'seniority' in the community, and the thematic areas in which they showed a specific interest. The two environments, while appearing closely structured and similar to those used for traditional online courses, did not require either compulsory attendance with regular internet contact, or a pre-ordered use of resources.

The Aladin 2.0 space

The learning space set up for the 'junior' teachers was divided into two sections: 'familiarisation with online resources' and 'laboratory of ideas'.

The first section was dedicated to activities of familiarisation with some of the best-known 2.0 resources. Here, the teachers were able to acquire familiarity with the tools which most interested them, consulting tutorials and articles, doing guided exercises, watching videos or consulting sites that presented experiences using those same tools in the school environment. These resources, prepared by ITD researchers in collaboration with the GDS experts, were used in self-learning mode. However, the teachers had at their disposal both a technological and a methodological/ educational support for the use of the proposed resources. This was provided by the researchers and the GDS, via forums or videoconferencing sessions.

The second section of the Aladin 2.0 environment, the 'laboratory of ideas', was conceived as a work and discussion space for the planning of experiments to be carried out in class with the students. The teachers were able to formulate and develop their own projects, sharing them with their peers and seeking support from the experts.

The Aladin 2.1 space

An environment was set up for the 'senior' members, with sections dedicated to some advanced training paths for the planning of projects on specific themes. The ITD researchers believed that it was through these paths that the 'senior' teachers would succeed in potentiating their skills in the educational use of 2.0 resources and in assessing how far the use of these tools might enrich and improve their teaching. Each section included a forum for discussion with colleagues and for support from experts on the specific chosen path, a folder with structured documentation for in-depth study of the chosen theme, and a reference guide on how to explore the resources which were made available.

The activities

The activities of the programme were divided up into the following stages (see Figure 7.2):

1. *Approach to Web 2.0 resources*: This stage preceded the preparation of projects as it was necessary for the teachers to examine the features of the tools which they knew only slightly or not at all, to learn their uses and understand the innovative teaching experiments in which they could be employed.
2. *Development of small short-term projects*: These lasted from two weeks (recommended for less expert teachers) to two months (for the teachers undergoing more advanced training). The projects could involve more than one teacher and be based on interdisciplinary themes, or be developed by just one teacher who wanted to pursue an educational goal linked to a topic from his/her own subject.
3. *Project revision*: This was directly organised by the ITD researchers. The project forms prepared by the teachers were read and commented on by the researchers, who assessed their clarity, coherence, feasibility and educational sustainability.

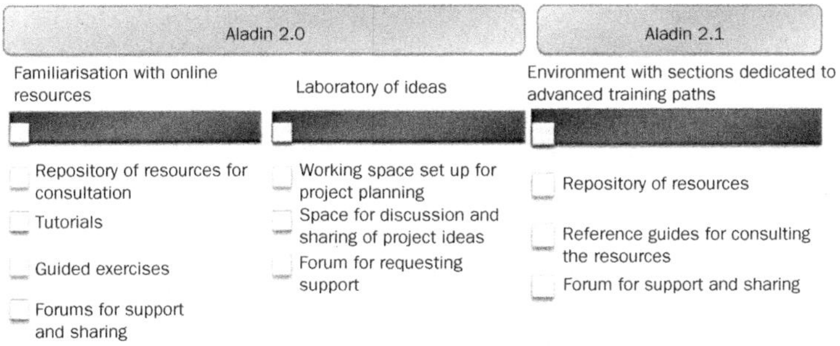

Figure 7.2 Scheme of the environments set up for the Aladin 2.0 and Aladin 2.1 programmes

4. *Trialling of the projects in class*: In this stage, the teachers worked independently with their own students. This situation amounted to a real test run, given the unpredictability of both the results and students' reactions. In this stage, the only support the teachers could rely on was provided online by the researchers, by their more expert colleagues (the 'seniors'), or by the GDS experts who in cases of real emergency were also ready to intervene on the spot.
5. *Assessment of the impact on teaching and learning*: The main aim of this stage was to work out approaches for assessing the effect produced by the introduction of ICTs into daily teaching practices. The methods adopted in the two Aladin programmes were different.

In Aladin 2.0, the process of assessment was performed via a kind of story-telling, i.e. inviting the teachers to express and share with colleagues their impressions of students' reactions to the proposed activities, as well as of the dynamics created, the difficulties encountered, and any unexpected 'discoveries'.

The work of participants in Aladin 2.1 was instead more structured. Participants were asked to collaborate with the ITD researchers in working out more exactly both the general approach to be adopted, and the tools for measuring the impact in terms of improvement in students' learning processes and performances, increase in participation and changes in attitude. We sought to encourage an educational and integrated assessment approach which would take into account not only the multiple dimensions of learning, but also the opinions expressed by all the actors involved, including the students themselves. Cross-checking of the data was expected to produce a more complete, pervasive assessment (Repetto, 2010) which would reveal those more embedded, latent elements that are difficult to capture with traditional assessment approaches.

The cycle of activity described above could be repeated by teachers at any point of the school year (Figure 7.3). Thus, they could plan and trial further projects in a wholly autonomous way, managing their own informal training in the acquisition of new skills and abilities.

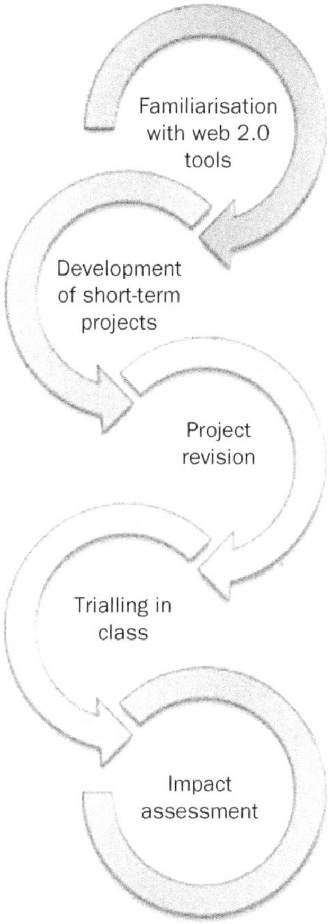

Figure 7.3 The iterative model with the stages of each Aladin programme

Parallel to this, the resources for familiarisation with Web 2.0 and its educational use were regularly updated, and new tools added. A kind of buffet of continuously-expanding resources was laid out on the Moodle platform, which could be dipped into on the basis of teachers' personal interests in the educational use of 2.0 resources.

The paths

The paths proposed within each programme, while complying with the general model of procedure, were however differentiated in accordance with the method of developing the educational project and the chosen theme.

The 'junior' teachers were offered:

(A) a path for those wanting to prepare a simple, independently-devised project;

(B) a path for those preferring to use a project idea already prepared by ITD researchers.

The 'senior' teachers were offered:

(C) a path for further investigation of 'impact assessment' – when planning the experience to be conducted in class, teachers were asked to define a set of criteria and tools for assessing the impact that the educational use of 2.0 resources could have on learning and teaching processes;

(D) a path on the theme of 'digital citizenship', in which the projects would aim at educating students on the one hand in the skilled use of web resources, and on the other in being responsible web 'citizens'.

The Aladin project: results and attestations

The Aladin teachers developed, trialled and documented 24 projects altogether. This count does not include the projects carried out independently, i.e. those whose results were not communicated to the project community, or those which were not completed.

The results

The majority of the teachers (three-quarters of the total number) opted for path A, while none of them chose to develop the pre-structured projects proposed in path B. This demonstrates that, although the number of projects in relation to the total number of teachers is rather low, ideas were not lacking and the educational objectives around which to develop the projects were quite clear. The remaining quarter of the teachers, the so-called 'seniors', were equally divided between paths C and D.

Regarding school level, the teachers were distributed more-or-less equally over the three levels, with slightly lower numbers for the upper secondary level.

As for the approach adopted in the projects, in 54 per cent of cases the activity proposed to pupils was of a collaborative type; in 33 per cent the mixed approach was chosen, where moments of individual work are alternated with group activities; and in 13 per cent the activity was exclusively individual.

Correlating the approach adopted with the school level, all the primary school teachers opted for the collaborative approach. This confirms the general situation in Italian primary schools, where group learning activities tend to be widely practised and more common than at the other school levels.

The lower secondary level teachers expressed no preference for any particular approach, while the most widely preferred approach for upper secondary teachers was the mixed one. Regardless of school level, the individual approach was chosen only for scientific subjects.

Considering the different subject areas, one-quarter of the projects regarded interdisciplinary themes, one-quarter scientific subjects, 17 per cent literature, and 8 per cent maths and English.

Overall, 42 per cent of the projects regarded the scientific field, 33 per cent the humanities, and 25 per cent multidisciplinary themes (Figure 7.4), which were mostly preferred in primary schools and a lot less at the other school levels.

No significant link was found between the subject area and the technological tools chosen. Correlating instead the chosen tool with the school level, it was found that test creation environments (HotPotatoes) were widely used in the lower secondary school projects, conceptual map creation tools were very popular in primary school projects, while no prevalence of any particular tool over another was found in the experiences proposed in the upper secondary school.

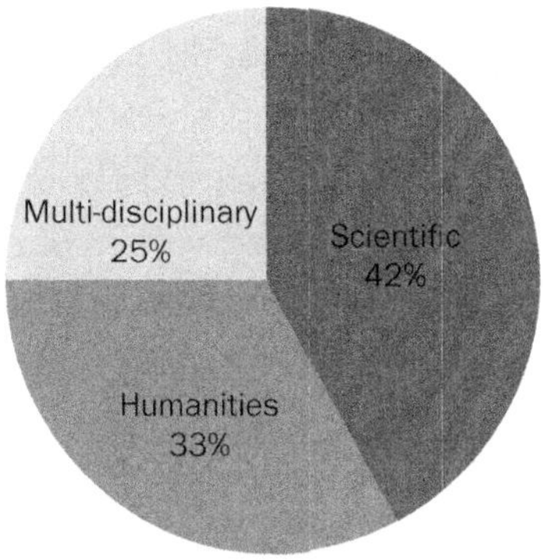

Figure 7.4 **Distribution of chosen topics in the Aladin projects**

Generally speaking, the most-used tool was the one for conceptual map construction (38 per cent), followed by test and quiz creation environments (25 per cent), the wiki tool (21 per cent) and the resources available on the Moodle platform (17 per cent) (see Figure 7.5). The other tools were all used in a smaller volume of projects (in the region of 4 to 8 per cent).

In 60 per cent of the projects, a single tool was adopted; in the remaining 40 per cent, the teachers opted to combine at least two technologies. The most commonly-recurring combinations were maps and wikis, maps and tests, and two or more resources from the Moodle platform.

The combined use of more than one technology, which denotes a better knowledge of online resources and their educational potential, was favoured in the projects promoted by a single teacher, above all in the secondary schools and for the scientific subjects.

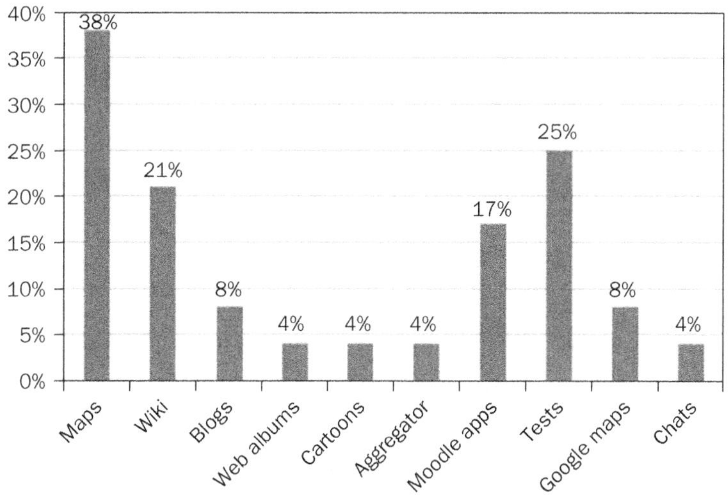

Figure 7.5 Tools adopted in the Aladin projects

Participants' attestations

Considering the results of the experimentation as expressed by the participants in the projects themselves, the overall balance was positive.

The students appreciated the activities proposed as a whole: using a new approach to a school subject and a new technology which in many cases is known only outside the school context, made learning of the proposed content more pleasant and interesting. Adopting more suitable educational strategies and choosing tools which are able to capture students' attention can also encourage them to tackle more complex content or what they consider to be less interesting topics.

The quotation below is a student's comment posted on a blog inside a learning unit dedicated to poetry:

> ...to fully understand poetry you must feel the emotions the poet wants to convey to us. Thus, in order to understand poetry and even to make poetry, you should transmit emotions and not keep them inside you. I completely agree with some of my classmates as well, who say that poetry is always the same because every year the same authors are proposed again ... but I'd also like to add that it depends how you tackle them. In the sense that it depends on how teacher suggests you should make poetry. This year for instance is becoming really interesting, as we're making poetry in a different way, not only with reviews and boring classroom explanations. We're making poetry with music, images, recordings...

The pupils of the first years of primary school, whose attitude towards the new technologies was even more favourable and spontaneous, also became excited by and involved in the suggested tasks, as a group of teachers attested:

> During the various phases of the work we particularly concentrated on the relational dynamics among the children, detecting good levels of collaboration and mutual help. The most restless pupils were able to work autonomously, with attention and concentration; the insecure ones worked tranquilly, and we didn't perceive that 'performance anxiety' which accompanies them during classroom activities. We are always astonished at how particularly receptive pupils using technology are, and how easily they are able to convey contents which in other circumstances take longer.

The 'senior' teachers who conducted an assessment of the impact of Web 2.0 use on students' learning discovered that the process of creating shared projects facilitates the development of transverse abilities such as relational, expressive and problem-solving skills, i.e. skills which are not subject-specific, but can be equally important and applicable in other contexts too.

The 'additional' assessments which students were asked to make with regard to their own contribution and that of their group enabled the teachers to discover aspects which would have been difficult to capture through mere observation of the group dynamics in the classroom, e.g. the change in certain attitudes, or the evolution of certain group dynamics.

Conclusion

Research argues that the e-maturity of schools in terms of provision, management and use of technology across the curricula (Butt and Cebulla, 2006), is usually associated with an efficient integration of technology into education (Lim et al., 2003; Somekh et al., 2007; Underwood et al., 2010). In Aladin, all schools were supported in the use of NMTs as considerable funds were allocated to investment in technologies and professional development courses. Nevertheless, the high e-maturity of the schools involved was an ideal condition but not such as to disrupt teachers' own personal teaching and learning practices. In order to integrate the use of 2.0 technologies into their teaching practices, teachers must evolve into 'e-teachers' (Trentin, 2013). Nevertheless, they are unable to do this without adequate training, also because in their subject-specific basic education they are rarely introduced to innovative teaching practices other than the traditional ones.

As the main objective of Aladin was the diffusion of pedagogical approaches centred on the use of 2.0 resources and hence on the peer collaboration between the main actors in the educational process (i.e. both teachers and students), a teacher training action based on those same principles was proposed. In other words, the chosen training strategy was not the organisation of a course (formal approach), but the diffusion among the teachers of the habit and mindset of being themselves the main actors in their own lifelong professional learning (informal approach). This was done by suggesting methods for using Web 2.0 resources and adopting new pedagogical approaches, and by offering constant online and onsite support from experts (researchers and experienced teachers).

The teachers' response was qualitatively very positive (the activities planned and trialled in the classroom were excellent), but quantitatively rather poor (only a low number of teachers completed the paths proposed in Aladin). The low level of adhesion and sporadic attendance of the teachers for the proposed activities revealed that teachers are not yet ready to accept training based on an informal learning process.

In Aladin, teachers perceived this type of training as optional, to be followed just occasionally. This may be because they are still anchored to the classical scheme of face-to-face classroom courses, with fixed timetables and places, a scheme that is in contrast to the use of technologies and methodologies based on the 2.0 philosophy. By adopting this attitude, teachers deny themselves the chance to enter into the collaborative spirit of a professional community, as well as failing to understand the overall significance of a type of education whose learning model is based on the active role of the participant.

If, in their training, teachers are not motivated to use technologies to manage their own learning process, it is difficult to imagine them being able to propose study activities based on the potential of 2.0 technologies. As Underwood claims in a Becta report, evidence shows that technologies moving teachers outside the 'comfort zone' tend to provoke high rejection rates, and that there is a strong reluctance to disrupt familiar practices. Research on teachers' response to the use of technologies in education has often identified teachers as a barrier instead of a force for change (Underwood and Dillon, 2011). Evidence also suggests that discontinuity between teachers and technology is in the very nature of those who decide to teach. Teachers as a group are low technology users as compared with the general population (Sandford et al., 2006). Thus, Underwood and Dillon claim that reluctance towards embedding technology into educational processes seems to be a natural characteristic of the teaching profession, although they also admit that a more positive picture of this profession is given by other research.

In any case, one of the key factors for innovating teaching practice is to create the belief that NMTs really can offer new possibilities both on the educational/pedagogical level and in acquiring transversal skills and abilities. Teachers should become aware that they are carrying out an educational process whose effects are all-embracing. In other words, they must perceive of this process as one which is not (or at least not only) strictly linked to the subject area, and also as a medium/long-term investment with a strong, positive impact on both students' personal and social spheres.

A multiplicity of other aspects, in fact, needs to be considered, such as education to digital citizenship; acquisition of collaborative work and study methods; and acquisition of skills in the use of online technologies to activate synergies aimed at problem-solving, the selective search for information, resources and knowledge sharing. We are talking about knowledge and skills which are now the necessary basis of any lifelong learning processes, when individuals need to be personally responsible for their own continuous updating throughout their professional career. Thus, these abilities may be considered life skills that the present young generations will need in order to be able to live effectively in tomorrow's society.

Instead, the type of knowledge and skills and the method presently used to impart them are far from students' real needs. While teachers are still trying to figure out how to support pupils with more personalised learning experiences, many students are already creating personalised learning environments for themselves outside school using NMTs, through which they are also acquiring 'academic' knowledge (Underwood and Banyard, 2010).

After a period in which formal education dominated, the widespread diffusion of NMTs is now shifting the emphasis onto informal learning (Banyard and Underwood, 2009).

I believe that, to bridge the gap between formal and informal learning, schools should assume the role of prime, if not unique, educational agency, by cooperating with other emerging agencies which are the sources of informal learning, and by systematically promoting initiatives like the one described in this chapter, in which teachers can experiment with 2.0 resources in the learning process in first person and informally, in order to be sufficiently confident when they are performing Web 2.0-centred learning activities with their students. Moreover, teachers should continuously reflect on their own freshly-acquired practices and learn from their colleagues. Their systematic and intersubjective reflection within communities of practice, also based on evidence regarding student and teacher performance should, Imants (2002) argues, lead to their reinterpreting their role and their duties in an active, conscious and exploratory manner.

Note

1. Bozen School Board Support Group for Education with ICTs.

Bibliography

Attwell, G. (2007) 'The personal learning environments – the future of eLearning?' *eLearning Papers* 2(1): 1–7.

Banyard, P. and Underwood, J. (2008) 'Understanding the learning space', *eLearning Papers* 9: 4–12.

Burbank, M.D. and Kauchak, D. (2003) 'An alternative model for professional development: investigations into effective collaboration', *Teaching and Teacher Education* 19: 499–514.

Butt, S. and Cebulla, A. (2006), *E-maturity and School Theoretical Concepts and Research. A Secondary Analysis of COL Evaluation Data*, London: National Centre for Social Research, Becta.

Condie, R., Munro, B., Seagraves, L. and Kenesson, S. (2007) *The Impact of ICT in Schools – A Landscape Review*, Coventry: Becta.

Eraut, M. (1994) *Developing Professional Knowledge and Competence*, London: Falmer Press.

Greenwood, D.J. and Levi, M. (1998) *Introduction to Action Research*, London: Sage Publications.

Hennessy, S., Ruthven, K. and Brindley, S. (2005) 'Teacher perspectives on integrating ICT into subject teaching: commitment, constraints, caution, and change', *Curriculum Studies* 37(2): 155–92.

Imants, J. (2002) 'Restructuring schools as a context for teacher learning', *International Journal of Educational Research* 37: 715–32.

Johnson, L., Adams, S. and Cummins, M. (2012) *NMC Horizon Report: 2012 K–12 Edition*, Austin, TX: The New Media Consortium.

Kennedy, A. (2005) 'Models of continuing professional development: a framework for analysis', *Journal of In-Service Education* 31(2): 235–50.

Lim, C.P., Khine, M.S., Hew, T., Wong, P., Shanti, D. and Lim, B. (2003) 'Exploring critical aspects of information technologies integration in Singapore schools', *Australian Journal of Educational Technology* 19: 1–24.

McKinney, S.J. and Wilson, A. (2008) 'Teachers, professionalism and continuing professional development', paper presented at Towards Professional Wisdom Conference, 26–28 March, Edinburgh.

Morris, D. (2010) 'E-confidence or incompetence: Are teachers ready to teach in the 21st century?' *World Journal on Educational Technology* 2(2), available at: *http://www.world-education-center.org/index.php/wjet/article/view/180* (accessed 6 November 2011).

OFSTED (2009) *The Importance of ICT: Information and Communication Technology in Primary and Secondary Schools, 2005/2008*, London: Ofsted.

Repetto, M. (2010) 'An approach for analysing interactions within virtual learning communities', in B.K. Daniel (ed.), *Handbook of Research on Methods and Techniques for Studying Virtual*

Communities: Paradigms and Phenomena, Hershey, PA: IGI Global Publications, pp. 654–70.

Richter, D., Kunter, M., Klusmann, U., Lüdtke, O. and Baumert, J. (2011) 'Professional development across the teaching career: teachers' uptake of formal and informal learning opportunities', *Teaching and Teacher Education* 27(1): 116–26.

Sandford, R., Uiksak, M., Facer, K. and Rudd, T. (2006) *Teaching with Games: Using Commercial Off-the-shelf Computer Games in Formal Education*, Bristol: Nesta Futurelab.

Somekh, B., Underwood, J., Convery, A., Dillon, G., Jarvis, J., Lewin, C. et al. (2007) *Evaluation of the ICT Test Bed Project: Final Report*, Coventry: Becta.

Trentin, G. (2010) *Network Collaborative Learning*, Oxford: Chandos Publishing.

Trentin, G. (2013) 'Network and mobile technologies in education: a call for e-teachers', in G. Trentin and M. Repetto (eds), *Using Network and Mobile Technology to Bridge Formal and Informal Learning*, Oxford: Chandos Publishing, pp. 153–82.

Trentin, G. and Repetto M. (2012) Becoming a networked teacher: a case study of the Aladin Project in Italy', in Q. Lin (ed.), *Advancement in Online Education: Exploring the Best Practices*, Vol. 1, Hauppauge, NY: Nova Science Publishers Inc, pp. 47–63.

Underwood, J. and Banyard, P. (2010) 'Self-regulated learning in technology enhanced learning environments in Europe: facilitators and barriers to future development', in R. Carneiro, P. Lefrere, K. Steffens and J. Underwood (eds), *Self-regulated Learning in Technology Enhanced Learning Environments. A European Perspective*, Rotterdam: Sense Publishers, pp. 155–63.

Underwood, J. and Dillon, G. (2011) 'Chasing dreams and recognising realities: teachers' responses to ICT', *Technology, Pedagogy and Education* 20: 343–56.

Underwood, J., Baguley, T., Banyard, P., Dillon, G., Farrington-Flint, L., Hayes, M. et al. (2010) *Understanding the Impact of technology: Learner and School-level Factors*, Coventry: Becta.

Wenger, E. (1998) *Communities of Practice: Learning, Meaning and Identity*, Cambridge: Cambridge University Press.

8

A conclusive thought: the opportunity to change education is, literally, at hand

Cathlee A. Norris and Elliot Soloway

Abstract: We have known for more years than we might care to admit what a classroom should look like in order for substantive learning to take place. But instead of a *learn-by-doing* (also known as *project-based learning*, *inquiry-based learning*, *active learning*, etc.) didactic pedagogy/direct instruction still rules the land. Whether it is a teacher at the front of the room expounding or a Khan Academy video expounding or an electronic whiteboard instead of a blackboard at the place of the expounding or if the students can immediately respond to expounding with a clicker – that is still direct instruction pedagogy and we all know that direct instruction pedagogy, while useful in places, ought not to be the overwhelmingly dominant pedagogy. So, why has this pedagogy not been replaced by the better pedagogy? The truth is this: enacting a learn-by-doing pedagogy (or project-based learning, inquiry-based learning, active learning, etc.) is really difficult to do; it requires considerable time on the teacher's part as well as deep content knowledge in order to tailor the learning experiences to the particular strengths and weaknesses of the student.

Keywords: learning-by-doing pedagogy, mobile computing, age of mobilism

Introduction

Professors, who devote their lives to what is, relatively speaking, a narrow area, and who teach one course per semester and have one to three PhD students can enact a learn-by doing pedagogy. Professional schools such as medicine or business, which need to provide a case-based learning environment, charge considerable sums and thus can bathe students in instructional support. Learn-by-doing is challenging to enact and without question it is expensive and it does not scale well at all.

But, if a K-12 student is lucky, he or she will experience a learn-by-doing pedagogy. Highly talented and motivated teachers – artisan teachers, as Tom Carroll[1] calls them – can and do create classrooms with an active learning pedagogy. Those teachers are the ones that are remembered by their students – remembered for having helped transform their lives in some significant way.

On the other hand, for the everyday teacher for whom teaching is a job not a calling, for whom working 16 hours a day – plus weekends – is not an option, enacting a learn-by-doing pedagogy with twentieth-century technology (i.e. pencil and paper and textbooks) is simply too hard to do. Indeed, the textbook publishers – not known for innovation or leadership – produce materials that support direct instruction that teachers readily use on a daily basis. And, with state-mandated testing, NCLB (no child left behind) and merit-based pay breathing down their necks, what is a teacher supposed to do? Tell the children what they need to know, tell them again louder and get them through those horrible tests.

We hasten to say this: we are in no way blaming teachers for this situation – far from it. We sympathize and empathize with teachers whose class sizes are increasing, who are taking on more parental-style responsibilities as children from dysfunctional homes bring that baggage into the classroom, and who regularly read in the newspapers the scorn heaped upon teachers and the teaching profession. At a recent political rally, one picketer had a sign that said: 'You are glorified baby sitters who leave work at 3 p.m. You deserve minimum wage'. Under these conditions, again, what is a classroom teacher to do?

Technology = opportunity

While technology can be just a bunch of gadgets, *good technology* is really *good opportunity* – good technology enables people to do things they literally couldn't do before. For example, Dell Computer didn't use the computer *to keep track of its inventory* of millions of parts like other PC manufacturers did. No, Dell used the computer and the internet to *eliminate inventory* – and in so doing, invented just-in-time PC manufacturing, scooping the industry and making big piles of money.

Company after company, industry after industry has leveraged those good gadgets – those good opportunities – and fundamentally changed the way they do business. Apple Computer is just Apple now, with half of its business coming from the iPhone; that transformation has happened in just four years. Oh, and Apple is the largest technology corporation in the world – all on the back of that miraculously-thin slab of aluminum-encased glass called the iPhone.

Sadly, K-12 missed the desktop revolution; it missed the internet revolution; and it missed the laptop revolution. Just having computers in your school does not mean your school leveraged the technology – leveraged the opportunity. For the most part, K-12 simply assimilated – integrated – desktops, the internet and laptops into the existing pedagogy; indeed, the electronic whiteboard and clickers just made direct instruction pedagogy even easier. The clicker (response pad) took off in K-12 when it came with a test bank of questions that teachers could use to create tests quickly and easily. Playing Jeopardy,[2] while certainly more fun than doing worksheets, is still, at its core, drill-and-practice; Jeopardy is an integral part of direct instruction, 'I teach', pedagogy.

But K-12 is not going to miss the mobile technologies revolution! K-12 will not be sitting out this opportunity – no way, no how!

The Age of Mobilism = opportunity for K-12 to finally change

Steve Jobs proclaimed that we are in the post-PC era. That's a negative way of naming our times. Calling it the Age of Mobilism more accurately reflects what mobile technologies are enabling. Mobile technologies are scaling faster than any technology ever. There are roughly 5.3 billion mobile phones and only 1.6 billion television sets. Pundits predict that feature phones – those 'simple' phones that are used to make phone calls – will not be available in two to three years as there is no money to be made in that business. Fifteen thousand new apps are published each week onto the smartphone marketplaces – 15,000 – per week. More smartphones are now being purchased than PCs; more time is spent using apps on a mobile device than is spent surfing the web. Indeed, we are moving at bullet train speed into the Age of Mobilism.

Mobile technologies are the opportunity that K-12 has been waiting for: just as a rising tide raises all boats, mobile technologies enable all teachers – not just the so-called artisan teachers – to move to a learn-by-doing pedagogy.

Simply put: mobile technologies, unlike twentieth-century paper and pencil technologies, will make it easier for all teachers to enact a learn-by-doing pedagogy. How? The mobile technologies of the twenty-first century will 'scaffold' learners, and help them take ownership and responsibility for their own learning. 'They [teachers] give the pupils something to do, not something to learn; and the doing is of such a nature as to demand thinking, or the intentional noting of connections; learning naturally results' (Dewey, 1916).

As we describe in more detail below, mobile technologies can scaffold the thinking that is demanded by a learn-by-doing pedagogy. Unlike twentieth-century paper-and-pencil technologies, mobile technologies can support students as they engage in 'the doing', as they engage in the 'noting of connections'. Now, because students can more easily engage in a learn-by-doing pedagogy, that in turn enables teachers – *all* teachers – to more easily engage in a learn-by-doing pedagogy.

Scaffolding the 'demanded thinking'

Imagine the following: in a classroom of 30 fifth-graders, each child has embedded in the palm of their hand an internet-connected computing device accessible essentially 24/7, inside the classroom and outside the classroom. In a learn-by-doing pedagogy, here are the key types of activities that such an arrangement enables:

- *Learn in context:* Science is full of definitions and otherwise abstract ideas. But in walking home from school, one young student notices an interesting root structure of an old, old tree. With the mobile learning device in his or her hand, snapping a picture of that root system is easy. What has just happened is that student, in Dewey-talk, has engaged in 'the intentional noting of connections' – she has linked a concrete image of a root system to an abstract idea discussed in the classroom. Moreover, the student has constructed an artifact, not just watched a video or listened to a lecture; the act of doing is the critically important element in learning.

- *Direct and immediate access to information, events, locations, data:* Following out this vignette, the student can use Google Goggles, an augmented reality app, to identify the tree, immediately, and provide additional information about the tree from the vast, internet store. This student doesn't need to wait until he or she arrives home to look up root system in a book – or even go online; with a device in his or her hand access to information – events, people, places, and things – is immediate and direct. Still further, information is but one element coming available over the internet. There are mounds of databases containing archival data (e.g. weather data, population data, pricing data) available freely; and, increasingly, as sensors proliferate, there is real-time data available on the internet. Also, events such as conferences, concerts, sales, etc. are directly accessible, as are places all over the globe that have an internet presence. The opportunity is this: if a student has an interest, no longer must their interest be mediated by a teacher or a textbook. Rather, the student can himself or herself immediately pursue to virtually any depth, their interest in a topic, an event, a location, etc. This level of direct access is unprecedented; the real challenge is making access ... more accessible!
- *Discuss, collaborate, and work as a team:* Still further, the young student brings that photo of the root system back into the classroom to show his or her classmates and then

a conversation naturally ensues. 'Wow, where did you get that picture? Look at the picture I took of a tree's root system. It looks different from your picture. Why?' Learning is in the conversation; learning occurs when there is discourse and discussion – in the context of information that can be immediately retrieved. What does a teacher do in this situation? Help the students learn; help the students understand what information to retrieve; work together, collaborate, to develop a deep, integrated understanding. In contrast to 'I teach', this is 'we learn'.

- *All the time, everywhere learning:* While in the past we have seen the classroom as the major place where schooling – and learning – takes place. But, in our 24/7 world, a child spends more time outside of school than in. While some use 'anytime, anywhere' learning, we feel that the term 'all the time, everywhere learning' better represents what children (and adults) do in their everyday lives. An internet-connected, mobile learning device is truly a cognitive prosthesis and supports 'all the time, everywhere learning'.
- *Mobile devices are not just computers:* While mobile devices can support computing activities such as text editing, spread-sheeting, etc., their functionality only *begins* there. These devices already come standard with GPS, accelerometer, camera, mic, ambient light sensor, and multi-touch screen. Tomorrow, these devices will come with all manner of sensors, e.g galvanic skin response, temperature, ECG, etc. The functionality that these sensors will unleash is almost beyond imagination.

In contrast, then to desktops on carts or desktops in a lab, or even many one-to-one laptop programs that use the computers as supplements to learning, as add-ons, as glorified typewriters or Google search engines, in the above description, the mobile learning device is being used as an essential tool in the learning enterprise, in the 'demanded thinking'. Indeed, in analyzing a number of empirical studies of one-to-one computer use, what we see is that it is precisely when the computers are used as essential tools – not just as supplements – that there are increases in student achievement – test scores go up.

In a truly learn-by-doing pedagogy, the learner is in fact taking on the ownership for his or her learning. The teacher is a mentor, a guide, that shapes and directs; that encourages and presses. But when the teacher walks away to work with another student, the mobile learning device is still there, serving as an essential scaffold for a student's learning. If a learner is supposed to take ownership for his or her learning, it is only reasonable that the learner be provided with tools that can support that responsibility. Mobile technologies, finally, provide learners with the opportunity to take ownership of their learning and thereby enable all teachers to engage in a learn-by-doing pedagogy. Mobile technologies are the key to scaling up learn-by-doing.

Barriers to change

While the opportunity is at hand, and while the need to change is great, there are still significant barriers that need to be overcome in order to take advantage of this exciting and important opportunity:

- *Lack of vision:* There are still the deniers – no, there is nothing wrong with the current system that a bit of tweaking won't correct; no, that technology-infused future is not needed in schools as that's not the way our parents learned; etc. Stay the course!
- *'But we don't have the money':* When a parent says 'we don't have the money for that' to their child, it really

means: 'we don't value that'. Ditto for schools. While there may not be sufficient funds to scale across the district, there is money to begin the movement to one-to-one, learn-by-doing – if only the district would value it.

As computing devices move from supporting enterprises to supporting individuals, the cost of those devices and the connection services, has come way, way down. But, for the 55 million K-12 school children in the USA – and the billions outside America – that cost is still too high – *much* too high. We estimate that an affordable price point for an internet mobile learning device for students – be they parent supported or school supported – is $10/month per student, where that cost includes a mobile learning device and an internet connection plan. That sounds low; but the telephone companies will still be making a profit at that price point.

BYOD (bring your own device) will happen; schools will stop banning mobile devices and allow students to bring their own device. While schools will no longer need to purchase the majority of devices, there are still costs involved in moving to one-to-one, e.g. networking, software, professional development, etc. BYOD will help; but, as always, there is no free lunch.

- *Curriculum, curriculum, curriculum:* Frankly, the biggest barrier is the lack of curriculum to support teachers as they transition outside their comfort zone. A well-articulated curriculum is needed by everyday teachers as they move from direct instruction to learn-by-doing. There is ample evidence that a thoroughly described curriculum helps teachers, especially the everyday teachers, to be more effective. Schools provide teachers with textbooks, guidebooks, professional development – for direct instruction pedagogy.

  But we have seen school after school move to one-to-one and provide teachers with precious little *curricular* support. Thus, it is not surprising that the teachers simply do as instructed and 'integrate' the technology into their direct instruction lesson plans. And what are the results? Correspondingly precious little gains in student achievement. We cannot expect to transition to one-to-one computing – be they laptops, netbooks, tablets, or smartphones – on the backs of the teachers; we cannot expect teachers, who already have full-time jobs, to recreate a whole new curriculum for a whole new pedagogy to leverage the affordances of one-to-one. What are we thinking?!

Waiting for the textbook companies – who are petrified that their fundamental business model is in the process of being totally disrupted – to develop the needed curriculum, is not a good idea. Rather, as we are seeing, districts are forming teams of teachers who, over the summer, are being paid to create new curriculum (why more districts do not band together is a mystery; the water cycle in Hoboken is the same water cycle in Newark, last we heard). Now, district-based curriculum development is not a long-term, scalable solution. But textbook companies, as they see that districts want new curriculum to support learn-by-doing, will produce it – eventually.

- *Lack of leadership:* Change is not going to be without bumps and pains. If the superintendent says: 'this is optional' then game over; teachers will fall back into their comfort zone. Change is *not* optional; that must be communicated and supported. Further, the local, state, and federal government must put pressure on the telephone companies to give K-12 special pricing. The precedent is there; the just-unveiled FCC (Federal Communication Commission) Connect to Compete Program provides poor families with internet access for $9.95/month. And the providers are still making a profit! It is time for educators to step up and be proactive, not reactive – for a change – and demand special school pricing from the telephone companies.

- *800 pound gorilla – assessment:* The emperor has no clothes and standardized tests drive instruction. As long as the tests are fact-based, then instruction will focus on the facts – especially as merit pay becomes more prevalent. While there are many positives in the Common Core Standards,[3] their focus is not on twenty-first century skills; in order to get everyone to agree, it is not surprising that the Common Core reflects basic, traditional – and common – educational values. Perhaps the new PISA (Program for International Student Assessment) tests, whose development is being driven by twenty-first century corporations such as Cisco, Microsoft, etc. will reset what is considered important to know. If we want a learn-by-doing pedagogy, then we need assessments that value what a learn-by-doing pedagogy enables.

Stay tuned!

For the record, we predict that by 2015, every child in every grade in every classroom in America will be using a mobile learning device for curricular purposes, 24/7. You can take that to the bank. The only issue is this: how will those devices be used? Will the barriers stand so that teachers will simply assimilate the mobile learning devices into their existing curriculum – with little or no impact on learning as the devices will be used as supplements, not as essential tools – or will schools overcome the barriers and move to a learn-by-doing pedagogy, where all the teachers are participating, and where students are being prepared for the knowledge-work, global marketplace? In the end, it is only a question of will: do we have it or not? Stay tuned!

Acknowledgment

Chapter 8 is a reprint by kind concession of Education Technology Publications of Norris C. and Soloway, E. (2012) 'The opportunity to change education is, literally, at hand', *Educational Technology Magazine* 52(2): 60–3.

Notes

1. Tom Carroll is President of the National Commission on Teaching and America's Future (NCTAF).
2. *Jeopardy* is an American television quiz show featuring trivia in history, literature, the arts, pop culture, science, sports, geography, wordplay and other topics.
3. The Common Core State Standards Initiative is a US education initiative that seeks to bring diverse state curricula into alignment with each other by following the principles of standards-based education reform.

Bibliography

Dewey, J. (1916) *Democracy and Education. An Introduction to the Philosophy of Education*, New York: Free Press.

Index

Printed in Dunstable, United Kingdom

91342993R00141